The End
of Time

The End of Time

The Next Revolution in Physics

Julian Barbour

OXFORD
UNIVERSITY PRESS

OXFORD
UNIVERSITY PRESS

Oxford   New York

Athens   Auckland   Bangkok   Bogotá   Buenos Aires
Cape Town   Chennai   Dar es Salaam   Delhi   Florence   Hong Kong   Istanbul
Karachi   Kolkata   Kuala Lumpur   Madrid   Melbourne   Mexico City   Mumbai
Nairobi   Paris   São Paulo   Shanghai   Singapore   Taipei   Tokyo   Toronto   Warsaw

and associated companies in

Berlin   Ibadan

Copyright © 1999 by Julian Barbour

First published by Oxford University Press, Inc., 1999

First issued as an Oxford University Press paperback, 2001
198 Madison Avenue, New York, New York 10016

Oxford is a registered trademark of Oxford University Press

Library of Congress Cataloging-in-Publication Data
Barbour, Julian B.
The end of time: the next revolution in our
understanding of the universe/by Julian Barbour.
p. cm.
Includes bibliographical references and index.
ISBN-13 978-0-19-511729-5; 978-0-19-514592-2 (pbk.)
ISBN 0-19-511729-8; 0-19-514592-5 (pbk.)
1. Space and time. 2. Relativity (Physics).
3. Quantum theory. I. Title.
QC173.59.S65 B374 2000
530.11 21—dc21   99-044319

7 9 10 8
Printed in the United States of America
on acid-free paper

CONTENTS

PART 4   QUANTUM MECHANICS AND QUANTUM COSMOLOGY

LIST OF DISPLAY ITEMS

THE STORY IN A NUTSHELL

Two views of the world clashed at the dawn of thought. In the great debate between the earliest Greek philosophers, Heraclitus argued for perpetual change, but Parmenides maintained there was neither time nor motion. Over the ages, few thinkers have taken Parmenides seriously, but I shall argue that Heraclitan flux, depicted nowhere more dramatically than in Turner's painting below, may well be nothing but a well-founded illusion. I shall take you to a prospect of the end of time. In fact, you see it in Turner's painting, which is static and has not changed since he painted it. It is an illusion of flux. Modern physics is beginning to suggest that all the motions of the whole universe are a similar illusion – that in this respect Nature is an even more consummate artist than Turner. This is the story of my book.

Snow Storm – Steamboat off a Harbour's Mouth Making Signals in Shallow Water, and Going by the Lead. The Author was in this Storm on the Night the Ariel *left Harwich* (1842). The 67-year-old Turner claimed that he had made the sailors bind him to the *Ariel's* mast so that he should be forced to experience the full fury of the storm.

PREFACE

We must philosophize about these things differently.
Johannes Kepler

On a beautiful October afternoon in 1963 I travelled to the Bavarian Alps with a student friend, Jürgen. We planned to spend the night in a hut and climb to the peak of the Watzmann at dawn next day. On the train, I read an article about Paul Dirac's attempt to unify Einstein's general theory of relativity with quantum theory. A single sentence in it was to transform my life: 'This result has led me to doubt how fundamental the four-dimensional requirement in physics is.' In other words, Dirac was doubting that most wonderful creation of twentieth-century physics: the fusion of space and time into space-time.

I never climbed the Watzmann. When Jürgen's alarm rang an hour before dawn, I awoke with a splitting headache. I still remember vividly the brilliant stars of Orion and the other winter constellations, high in the sky before the October dawn. But stars or no stars, I could not face the climb with that headache. Jürgen set off alone, but I took two aspirin and went back to my bunk. Waking an hour or two later, I fell to thinking about Dirac's words. Might the notion of space-time have been a mistake? This prompted an even more fundamental question: *what is time?* Before Jürgen had returned, I was – and still am – the prisoner of this question.

Richard Feynman once quipped, 'Time is what happens when nothing else does.' My conclusion, reached within a few days, was the exact opposite: time is nothing but change. I spent hours and hours pacing through the Englischer Garten in Munich while persuading myself of this fact. Physics must be recast on a new foundation in which change is the measure of time, not time the measure of change. After a week or two I had become so gripped by the issue of time that I decided to go to Cambridge, where Dirac was the Lucasian Professor, as Newton had once been, to try to explain my ideas to him. I could have saved myself the trouble. Although I had studied mathematics in Cambridge from 1958 to 1961, I had never been to Dirac's lectures and did not know he was a man of few words, seldom engaging in general discussions even with his most distinguished colleagues. I did speak to him briefly on the telephone, on which he introduced himself with 'This is Mr Dirac', but, very reasonably I am sure, he ended the conversation quite quickly.

If the trip to Cambridge failed in that way, it had for me a most fortunate side effect. While over in England, I went back to the village in north Oxfordshire where I had grown up. There I found my younger brother wondering if he could find the money to become a farmer. New College in Oxford had suddenly decided to auction, in separate lots, a farm they owned in the village. The auction was set for 21 November. We had both received some money from my father as a way of avoiding death duties. When it became clear, just twenty-four hours before the auction, that my brother might manage to get together, with loans, enough money to buy the land but not the farmhouse and buildings, I decided on the spur of the moment to bid for them myself and rent them to my brother for a few years until he could make other arrangements. At dawn on the morning of the sale, my father woke me and my mother woke my brother. They had been unable to sleep, lying awake all night worrying about our plans. We must give up the idea. However, our bank manager encouraged us to go ahead. A few hours later – and about twenty-eight hours before President Kennedy was assassinated – I was the proud owner of College Farm, as it was and still is called. Built at the end of the Commonwealth in 1659 and standing next to a fine medieval church, it is one of the best-preserved yeoman farmhouses in the country.

I have told you this story of my serendipitous purchase because it significantly affected the kind of scientist I became. After I returned to Munich, and my brother started to plough his fields, I decided to give up the Ph.D. in astrophysics I had begun and turn to fundamental physics, above all time. I did a Ph.D. on Einstein's theory of gravitation in Cologne and then started to think about a university position in Britain. But even then there was pressure to 'publish or perish'. If you could not turn out one or two research papers each year (now, crazily, one is expected to produce four or five) and do all the teaching and administrative duties, I was warned, you could not look forward to much of a career. But I might want to spend years thinking about basic issues before publishing anything. As luck would have it, I had learned Russian as a hobby while in Munich and had earned some money by translating Russian scientific journals. Once you get into such work, it goes quite fast, especially if you can dictate it. So I decided to earn my living that way, and work away at the question of time as and when I could. In 1969 my German wife and I moved into College Farm with our two small children, who were soon followed by two more. For twenty-eight years, until I felt the size of my pension fund allowed me to stop, I turned out translations at the rate of two and a half million words a year. I think that together they would fill about twenty metres of library shelves.

It was a great way to bring up a family, but an unconventional way to do physics. For years I never met anyone else at conferences who was not at a university or research institute. Now you do meet a few. James Lovelock, the creator of the Gaia theory of life on Earth, is a great advocate for going independent. I think it worked for me. I greatly valued the feeling that I could do just what I wanted when I wanted. Publication of papers led to fruitful collaboration with other physicists and trips to many parts of the world. I had the luxury of being able to work on topics other physicists felt they could not risk, either because nothing might come of the work or because their reputations would suffer. But they still liked to talk about them, and I made several good friends in this way. And all the while, I did seem to make some progress on the enigma of time. Key ideas came every five or six years, the most radical in 1991. In fact, 35 years on from that failed attempt on the Watzmann, I now believe that time does not exist at all, and that motion itself is pure illusion. What is more, I believe there is quite strong support in physics for this view. I have a vision and I want to tell you about it.

You may wonder how I can preface a belief that time does not exist with a bit of personal history. How can history be if there is no time? That is the great question, and my answer comes at the end of the book. Most of the book is about what evidence physics can offer for and against the existence of time. However, in the first part I try to explain, in the simplest terms possible, the main issues, and to relate them to your direct experience of time. I want to try to make sure, if you have bought or borrowed this book, that you do not put it down in despair, unable to understand what I am driving at. I hope also that this introduction will encourage you to read on to the details. Many are fascinating in their own right. Because temporal concepts are so deeply lodged in our experience and language, I shall often write as if time existed in the way most people think it does. The same applies to motion. Please do not think I am being inconsistent – I should have to use many more words to express everything in a timeless fashion.

I have tried to make the text self-contained and accessible to any reader fascinated by time. If you find some parts harder then others, please do not worry if you have to give up on them. Several non-scientists who read a much more technical first draft found they could simply skim the harder parts and still pick up much of the message. For this reason, the more technical material that is not completely central to the story is generally put in boxes – take that as a sign not to worry if you have difficulty digesting it (though I hope you will at least try it). Also, various digressions, of potential interest to all readers, and genuinely technical material for cognoscenti are to be found in the notes at the end. I suggest you look at them after you have read each chapter. To help readers with little or no scientific background, the most important technical terms appear in the Index so that you can readily locate explanations of them in the text if necessary. Books for further reading are also recommended.

I dedicated my first book to my wife and our children. I dedicate *The End of Time* to my indomitable mother, just ninety-six and still hearing the larks clearly and singing as lustily in her church choir. I dedicate it equally to the memory of my father, who died three years after my wife and I moved into College Farm. My father, whom I missed very much, had a most useful saying that I should like to share with you: 'Never believe anything anyone ever tells you without checking again and again.' That has saved me from many a disaster. A very good friend of mine, Michael Purser, once remarked that if my mother was the irresistable force, my father was surely the immovable object. Whatever the truth, I should not be here but for them. Being here is the supreme gift.

J.B.

South Newington, March 1999

Note This printing of the book differs from the initial hardback in the correction of some minor errors and misprints, additions to the bibliography and books recommended for further reading, and slight rearrangement of the Notes to take into account new results obtained with Niall Ó Murchadha after the book had been written. This recent work should, if it stands up to critical examination, strengthen my arguments that time does not exist. See especially p. 358.

ACKNOWLEDGEMENTS

Several people who have helped me greatly are mentioned in the text and notes, where it seemed more appropriate to express my gratitude to them. All of them also helped by reading some or all of an early draft and making comments. I am also grateful to several others (listed here in no particular order) who did the same: Dr Tiffany Stern, Michael Pawley, David Rizzo, Mark Smith, Dr Fotini Markopoulou, Gretchen Mills Kubasiak (with particularly detailed and helpful comments), Oliver Pooley, Dr Joy Christian, Cyril Aydon, Dr John Purser, Jason Semitecolos, Todd Heywood, John Wheeler (this is not J.A. Wheeler, though he did read the later draft, for which I am most grateful), Christopher Richards, Michael Ives, Elizabeth Davis and Ian Phelps. Joyce Aydon, Mark Smith and Tina Smith helped greatly with the preparation of the text. I should like to thank too Steve Farrar and his editor Tim Kelsey, who went to great trouble to report my ideas accurately in an article (entitled 'Time's assassin'!) in the *Sunday Times* in October 1998.

I am especially indebted to my friend Dierck Liebscher of the Astrophysikalisches Institut Potsdam, who prepared all the computer-generated diagrams (and also made helpful comments on the text).

Both my editors (Peter Tallack for the UK edition, Kirk Jensen for the North American edition) have done very well what editors of a book like this should do: be supportive but insist that it is for the popular market, not an academic text. It is not for me to judge how readable the final result is, but to the extent that it is, my readers must be grateful to them, as I am. I am also grateful to my copy-editor, John Woodruff, for numerous stylistic improvements and his thorough work. Lee Smolin, who appears often in the main text, needs to be mentioned especially here too, since he made the most valuable suggestion that I write the introductory chapters that comprise Part 1. Without these, the book in its first draft was much tougher.

My wife, Verena, and our children have been wonderfully supportive.

I also want to thank here my literary agent Katinka Matson and her partner John Brockman, founder of Brockman, Inc., not only for finding me quite the best publishers and editors I could hope for but also for a remark of John's that encouraged me to write the kind of book this has become. According to John, 'Roger Penrose has found the right way to write popular science today. He's really writing for his colleagues, but he is letting the public look over his shoulder.' For myself, I have certainly tried to write primarily for the general reader, but, in a reversal of John's aphorism, I shall be more than happy if my colleagues look over my shoulder. This is a serious book, and it draws its inspiration from the way Penrose's *The Emperor's New Mind* engages with intensity – passion, even – both the interested public and working scientists. That is what gives his book its cutting edge and thereby makes it more absorbing for the non-specialist. Richard Dawkins's *The Selfish Gene* is another example that comes to mind.

I have left to the end one other important person – you, the reader. As you will know from the Preface, I have tried throughout my life to fund my own research and would like to continue to do so. Every copy of this book that is bought (and borrowed from a library) helps me in this way. Thank you, and I do hope you get some pleasure from this book. I have enjoyed writing it. I hope to continue popularizing the study of time and will post details on my Website (www.julianbarbour.com) together with any significant developments of which I become aware in the study of time.

PART 1

The Big Picture in Simple Terms

As explained in the Preface, I start with three chapters in which I have attempted to present my main ideas with the minimum of technical details. The main aim is to introduce a definite way of thinking about *instants of time* without having to suppose that they belong to something that flows relentlessly forward. I regard instants of time as real things, identifying them with possible instantaneous arrangements of all the things in the universe. They are configurations of the universe. In themselves, these configurations are perfectly static and timeless. But how and why can something static and timeless be experienced as intensely dynamic and temporal?

That is what I hope to explain in simple terms in these first three chapters.

CHAPTER 1

The Main Puzzles

THE NEXT REVOLUTION IN PHYSICS

Nothing is more mysterious and elusive than time. It seems to be the most powerful force in the universe, carrying us inexorably from birth to death. But what exactly is it? St Augustine, who died in AD 430, summed up the problem thus: 'If nobody asks me, I know what time is, but if I am asked then I am at a loss what to say.' All agree that time is associated with change, growth and decay, but is it more than this? Questions abound. Does time move forward, bringing into being an ever-changing present? Does the past still exist? Where is the past? Is the future already predetermined, sitting here waiting for us though we know not what it is? All these questions will be addressed in this book, but the biggest remains the one St Augustine could not answer: what is time?

Curiously, physicists have tended not to ask this question, preferring to leave it to philosophers. The reason is probably the colossal and dominating influence of Isaac Newton and Albert Einstein. They shaped the way physicists think about space, time and motion. Each created a representation of the world of unsurpassed clarity. But having seen their way to a structure of things, they did not bother unduly about its foundations. This creates potential for confusion. Without question, their theories contain wonderful truths, but they both take time as given. It is a building block on a par with space, a primary substance. In fact, Einstein fused it with three-dimensional space to make four-dimensional space-time. This was one of the great revolutions of physics (Box 1).

BOX 1   The Great Revolutions of Physics

1543: The Copernican Revolution. In *On the Revolutions of the Celestial Spheres*, Nicolaus Copernicus proposed that the Earth moves around the centre of the universe. The modern meaning of revolution derives from his title. He established the form of the solar system. Curiously, the Sun plays little part in his scheme; he merely placed it near the centre of the universe. About sixty years later Johannes Kepler showed that the Sun is the true centre of the solar system, and with Galileo Galilei he prepared the way for the next revolution.

1687: The Newtonian Revolution. In *The Mathematical Principles of Natural Philosophy*, Newton formulated his three famous laws of motion and the theory of universal gravitation. He showed that all bodies – terrestrial and celestial – obey the same laws, and thus set up the first scheme capable of describing the entire universe as a unified whole. Newton created the science of mechanics, now often called dynamics, which ushered in the modern scientific age. He claimed that all motions take place in an infinite, immovable, absolute space and that time too is absolute and 'flows uniformly without relation to anything external'.

1905: The Special Theory of Relativity. In a relatively short paper on electromagnetism, Einstein showed that simultaneity cannot be defined absolutely at spatially separated points, and that space and time are inextricably linked together. What appears as space and what appears as time depends on the motion of the observer. He made startling predictions about the behaviour of measuring rods and clocks, and found his famous equation $E = mc^2$. In 1908 Hermann Minkowski formalized the notion of space-time as a rigid, indissoluble, four-dimensional arena of world events.

1915: The General Theory of Relativity. The special theory of relativity describes a world without gravitation. After an eight-year gestation, Einstein finally formulated his general theory of relativity in which the rigid arena of Minkowski's space-time is made flexible, responding to the presence of matter in it. Gravity is given a brilliantly original interpretation as an effect of the curving of space-time. The theory showed that time can have a beginning (the Big Bang) and that the universe can expand or contract. Although to a remarkable degree it was a creation of pure thought, many predictions of this theory have now been very well confirmed. It describes the large-scale properties of matter and the universe as a whole.

1925/6: Quantum Mechanics. This gets its name because it shows that some mechanical quantities are found in nature only in multiples of discrete units called quanta. This is a distinctive difference from the theories of Newton and Einstein, which are now called classical (as opposed to quantum) theories. The first quantum effects were discovered and described on an ad hoc basis by Max Planck (1900), Einstein (1905) and Niels Bohr (1913), while a consistent quantum theory was found in two different but equivalent forms: matrix mechanics, by Werner Heisenberg (1925), and wave mechanics, by Erwin Schrödinger (1926). Paul Dirac also made outstanding contributions. Quantum mechanics describes the properties of light, especially lasers, and the microscopic world of atoms and molecules. It is the bedrock of all modern electronic technology, but its results are bafflingly counter-intuitive and raise profound issues about the nature of reality. It is also puzzling that theories of completely different structures are used to describe the macroscopic universe (classical general relativity) and microscopic atoms (quantum mechanics).

Revolutions are what make physics such a fascinating science. Every now and then a totally new perspective is opened up. But it is not that we close the shutters on one window, open them on another, and find ourselves looking out in wonder on a brand-new landscape. The old insights are retained within the new picture. A better metaphor of physics is mountaineering: the higher we climb, the more comprehensive the view. Each new vantage point yields a better understanding of the interconnection of things. What is more, gradual accumulation of understanding is punctuated by sudden and startling enlargements of the horizon, as when we reach the brow of a hill and see things never conceived of in the ascent. Once we have found our bearings in the new landscape, our path to the most recently attained summit is laid bare and takes its honourable place in the new world.

Today, physicists confidently, indeed impatiently, await the next revolution. But what will it be? In 1979, when, like Newton and Dirac before him, Stephen Hawking became the Lucasian Professor at Cambridge, he announced in his inaugural address the imminent end of physics. Within twenty years physicists would possess a theory of everything, created by a double unification: of all the forces of nature, and of Einstein's general theory of relativity with quantum mechanics. Physicists would then know all the inner secrets of existence, and it would merely remain to work out the consequences.

Neither unification has yet happened, though one or both certainly could. (Hawking has recently said that his prediction still stands but that 'the twenty years starts now'.) For myself, I doubt that would spell the end of physics. But unification of general relativity and quantum mechanics may well spell the *end of time*. By this, I mean that it will cease to have a role in the foundations of physics. We shall come to see that time does not exist. Though still only a prospect on the horizon, this, I think, could well be the next revolution. What a denouement if it is!

I believe that the basic elements of this potential revolution – the reasons for it and its likely outcome – can already be discerned. In fact, as we shall shortly see, clear hints that time may not exist, and that quantum gravity – the unification of general relativity and quantum mechanics – will yield a static picture of the quantum universe, started to emerge about thirty years ago, but made remarkably little impact. This is one of my reasons for writing this book: these things should be better known. They are only just beginning to be mentioned in books for the general reader, and even most working physicists know little or nothing about them.

No doubt many people will dismiss the suggestion that time may not exist as nonsense. I am not denying the powerful phenomenon we call time. But is it what it seems to be? After all, the Earth seems to be flat. I believe the true phenomenon is so different that, presented to you as I think it is without any mention of the word 'time', it would not occur to you to call it that.

If time is removed from the foundations of physics, we shall not all suddenly feel that the flow of time has ceased. On the contrary, new time-less principles will explain why we *do* feel that time flows. The pattern of the first great revolution will be repeated. Copernicus, Galileo and Kepler taught us that the Earth moves and rotates while the heavens stand still, but this did not change by one iota our direct perception that the heavens do move and that the Earth does not budge. Our grasp of the intercon-nection of things was, however, eventually changed out of recognition in ways that were impossible to foresee. Now I think we must, in an ironic twist to the Copernican revolution, go further, to a deeper reality in which nothing at all, neither heavens nor Earth, moves. Stillness reigns.

People often ask me what are the implications of the non-existence of time. What will it mean for everyday life? I think we cannot say. Copernicus had no inkling of what Newton (let alone Einstein) would find, though it all flowed from his revolution. But we can be certain that our ideas about time, causality and origins will be transformed. At the personal level, thinking about these things has persuaded me that we should cherish the present. That certainly exists, and is perhaps even more wonderful than we realize. *Carpe diem* – seize the day. I expand on this in the Epilogue.

THE ULTIMATE THINGS

This book revolves around three questions: What is time? What is change? What is the plan of the universe? The only way to answer them is to examine the structure of our most successful theories. We must fathom the architecture of nature. What part, if any, is played by time in these theories? Can we identify the ultimate arena of the world?

These questions were forced upon physicists by the work I mentioned in the Preface. It is one of the two big (and almost certainly intimately connected) mysteries of modern physics (Box 2). Both are aspects of an as yet unbridged chasm between classical and quantum physics.

BOX 2   The Two Big Mysteries

As explained in Box 1, physicists currently describe the world by means of two very different theories. Large things are described by classical physics, small things by quantum physics. There are two problems with this picture.

First, general relativity, Einstein's theory of gravity, seems to be incompatible with the principles of quantum mechanics in a way that Newtonian dynamics and the theory of electromagnetism, developed by Michael Faraday and James Clerk Maxwell in the nineteenth century, are not. For these theories, it proved possible to transform them, by a process known as quantization, from classical into quantum theories. Attempts to apply the same process to general relativity and create *quantum gravity* failed. It was this technical work, by Dirac and others, which brought to the fore all the problems about time with which this book is concerned.

The second mystery is the relationship between quantum and classical physics. It seems that quantum physics is more fundamental and ought to apply to large objects, even the universe. There ought to be a quantum theory of the universe: *quantum cosmology*. But quantum physics does not yet exist in such a form. And its present form is very mysterious. Part of it seems to describe the actual behaviour of atoms, molecules and radiation, but another part consists of rather strange rules that act at the interface between the microscopic and macroscopic worlds. Indeed, the very existence of a seemingly unique universe is a great puzzle within the framework of quantum mechanics. This is very unsatisfactory, since physicists have a deep faith in the unity of nature. Because general relativity is simultaneously a theory of gravity and the large-scale structure of the universe, the creation of quantum cosmology will certainly require the solution of the only slightly narrower problem of quantum gravity.

One of the themes of my book is that this chasm has arisen because physicists have deep-rooted but false ideas about the nature of space, time and things. Preconceptions obscure the true nature of the world. Physicists are using too many concepts. They assume that there are many things, and that these things move in a great invisible framework of space and time.

A radical alternative put forward by Newton's rival Leibniz provides my central idea. The world is to be understood, not in the dualistic terms of atoms (things of one kind) that move in the framework and container of space and time (another quite different kind of thing), but in terms of more fundamental entities that fuse space and matter into the single notion of a possible arrangement, or configuration, of the entire universe. Such configurations, which can be fabulously richly structured, are the ultimate things. There are infinitely many of them; they are all different instances of a common principle of construction; and they are all, in my view, the different *instants of time*. In fact, many people who have written about time have conceived of instants of time in a somewhat similar way, and have called them 'nows'. Since I make the concept more precise and put it at the heart of my theory of time, I shall call them *Nows*. The world is made of Nows.

Space and time in their previous role as the stage of the world are redundant. There is no container. The world does not *contain* things, it *is* things. These things are Nows that, so to speak, hover in nothing. Newtonian physics, Einstein's relativity and quantum mechanics will all be seen to do different things with the Nows. They arrange them in different ways. What is more, the rules that govern the universe as a whole leave imprints on what we find around us. These local imprints, which physicists take as the fundamental laws of nature, reveal few hints of their origin in a deeper scheme of things. The attempt to understand the universe as a whole by 'stringing together' these local imprints without a grasp of their origin must give a false picture. It will be the flat Earth writ large. My aim is to show how the local imprints can arise from a deeper reality, how *a theory of time* emerges from timelessness. The task is not to study time, but to show how nature creates the impression of time.

It is an ambitious task. How can a static universe appear so dynamic? How is it possible to watch the flashing colours of the kingfisher in flight and say there is no motion? If you read to the end, you will find that I do propose an answer. I make no claim that it is definitely right – choices must be made, and many physicists would not make mine. If all were clear, I should not have promised *a* but *the* theory of time. In order not to interrupt the flow of the text, I make few references to the problems in my timeless description of the world. Instead, I have collected together all those of which I am aware in the Notes. Although, as will be evident throughout the book, I do believe rather strongly in the theory I propose, there is a sense in which even clear disproof of my theory would be exciting for me. The problems of time are very deep. Clear proof that I am wrong would certainly mark a significant advance in our understanding of time. In a way, I cannot lose! Whatever the outcome, I shall be more than happy if this book gives you a novel way of thinking about time, exposes you to some of the mysteries of the universe, and encourages even one reader to embark, as I did 35 years ago, on a study of time.

For the study of time is not just that – it is the study of everything.

GETTING TO GRIPS WITH ELUSIVE TIME

The hardest thing of all is to find a black cat in a dark room,
especially if there is no cat.
Confucius

We must begin by trying to agree what time is. The problems start already, as St Augustine found. Nearly everybody would agree that time is experienced as something linear. It seems to move forward relentlessly, through instants strung out continuously on a line. We ride on an ever-changing Now like passengers on a train. Each point on the line is a new instant. But is time moving forward – and if so through what – or are we moving forward through time? It is all very puzzling, and philosophers have got into interminable arguments. I shall not attempt to sort them out, since I do not think it would get us anywhere. The trouble with time is its invisibility. We shall never agree unless we can talk about something we can see and grasp.

I think it is more fruitful to try to agree on what an instant of time is like. I suggest it is like a 'three-dimensional snapshot'. In any instant, we see objects in definite positions. Snapshots confirm our impression; artists were painting pictures that look like snapshots long before cameras were invented. This does seem to be a natural way to think about the experience of an instant. We also have evidence from the other senses. I feel an itch at the same time as seeing a moving object in a certain position. All the things I see, hear, smell and taste are knit together in a whole. 'Knitting together' seems to me the defining property of an instant. It gives it a unity.

The three-dimensional snapshots I have in mind could be constructed if many different people took ordinary two-dimensional snapshots of a scene at the same instant. Comparison of the information in them makes it possible to build up a three-dimensional picture of the world in that instant. That is what I mean by a Now. It is very remarkable that such completely different two-dimensional pictures can be reconciled in a three-dimensional representation. The possibility of this ordering is what leads us to say that things exist in three-dimensional space. It leads to an even deeper 'knitting together' over and above the directly experienced sense of being aware of many different things at once (it is this that enables us to know instantly that we are seeing, say, six distinct objects without counting them individually). I regard space as a 'glue', or a set of rules, that binds things together. It is a plurality within a deep unity, and it makes a Now.

You may object that no experience is instantaneous, just as snapshots require finite exposures. True, but we can still liken instants to snapshots. It is the best idealization I know. It allows us to begin to get our hands on time, which is otherwise for ever slipping through our fingers. As instants, rather than an invisible river, time becomes concrete. We can pore over photographs, looking for evidence in them like military intelligence analysts studying satellite pictures. We can imagine 'photographing' our successive experiences, obtaining innumerable snapshots. Using them, we can identify the most important properties of experienced time.

THE PROPERTIES OF EXPERIENCED TIME

Suppose that the snapshots are taken when we are witnessing lots of things happening, say people streaming past us in a street, and that the snapshots (either two-dimensional, as directly experienced, or 'three-dimensional', as explained above), once taken, are jumbled up in a heap. A different person, given the heap, could relatively easily, by examining the details in the snapshots, arrange them in the order they were experienced. A movie can be reassembled from its individual frames. My notion of time depends crucially on the details that the 'snapshots' carry. It requires the richly structured world we do experience.

This imaginary exercise brings out the most important property of experienced time: its instants can all be laid out in a row. They come in a linear sequence. This is a very strong impression. It is created not by invisible time, but by concrete things.

It is harder to pin down other properties. I have already mentioned the difficulty of saying precisely what the powerful impression of moving forward in time consists of. We also have the intuition of length of time, or duration. Indeed, seconds, minutes, hours dominate our age, though you may not know how these precise notions have arisen. That is an important issue. Finally, there is the remarkably strong sense that time has a direction. A line traced in the sand does not by itself define a direction. If time is a line, it is a special one.

The evidence for time's direction is in the 'snapshots'. Many contain memories of other snapshots. We can do a test on time. We can stop at one of our experienced instants laid out in a line, and see that it contains a memory. We locate the remembered instant somewhere in the line. That defines a direction – from it to the memory of it. We can do this with other pairs of instants. They always define the same direction. Many other phenomena define a direction. Coffee cools down unless we put it in the microwave; it never heats up. Cups shatter when we drop them; shards never reassemble themselves and leap back up onto the table as a whole cup. All these phenomena, like memories, define a direction in time, and they all point the same way. Time has an arrow.

Thus, experienced time is linear, it can be measured and it has an arrow. These are not properties of an invisible river: they belong to concrete instants. Everything we know about time is garnered from them. Time is inferred from things.

NEWTON'S CONCEPTS

In 1687, Newton created precise notions of space, time and motion. Despite major revisions, much of his scheme remains intact. It is still close to the way many people, including scientists, think about time.

Newton's time is absolute. It flows with perfect uniformity for ever and nothing in the world affects its flow. Space, too, is absolute. Newton conceived of space as a limitless container. It stretches from infinity to infinity like a translucent block of glass, through which, nevertheless, objects can move unhindered. Space is a huge arena; time is a clock in the grandstand. Both are more fundamental than things. Newton could imagine an empty world but not a world without space and time. Many philosophers have agreed with him. So does the proverbial man in the pub, convinced that space goes on for ever and that 'there must have been time before the Big Bang'.

At any instant, all the things in the Newtonian world are at definite positions. His absolute space performs two distinct roles. As in the discussion above, it binds, or holds, things together, in one instant. But it also places them in a container. Imagine taking two-dimensional snapshots of a table in a room. Paint out the background room, and you could still reconstruct the form of the three-dimensional table, but you would not know where to place it. Newton insisted that the things in the world in any instant have a definite place, and he posited absolute space as a kind of room to provide that place. His fixed container persists through time. We could take real snapshots of the things in the world (Figure 1). Ideally, these snapshots should be three-dimensional, like space, and show all things relative to each other and their positions in absolute space, just as snapshots of a soccer match show the players, ball and referee on the pitch with its markings. The grandstand clock records the time.

According to Newton, all bodies move through absolute space in accordance with definite laws of motion which govern the speed and direction of the bodies in that space as measured by absolute time. The laws are

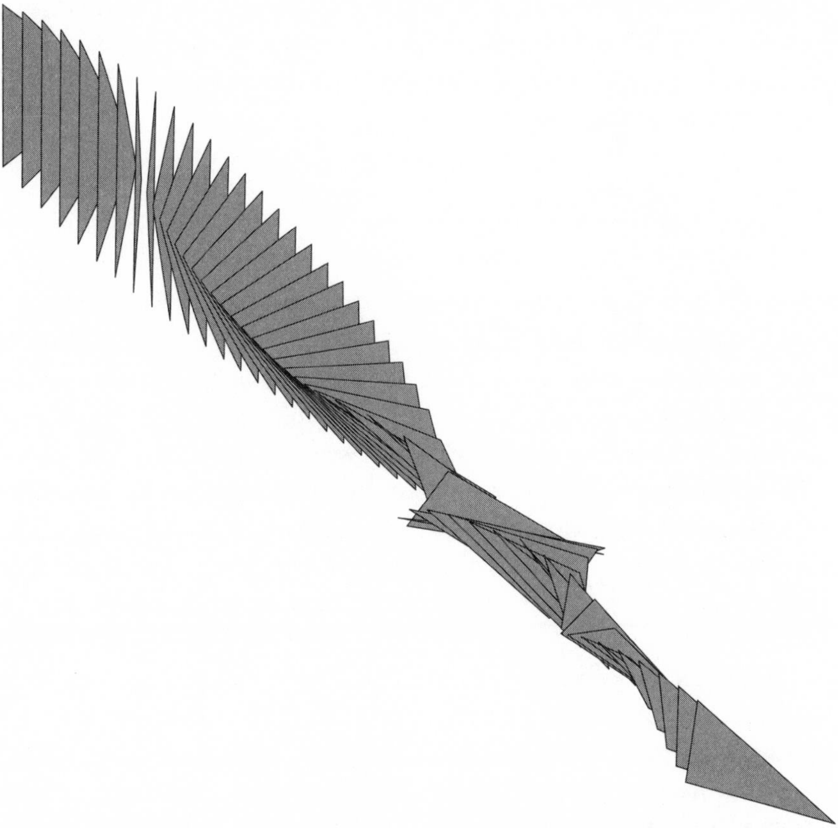

Figure 1 As explained in the text, Newton conceived of space as a container, or arena, and time as a uniform flow. The difficulty is that both are invisible. This diagram attempts to represent the way he thought about space and time. The blank white of the page is a two-dimensional substitute for the invisible three-dimensional space, and the effect of the flow of time is mimicked by supposing that it triggers light flashes at closely spaced equal intervals of time. These flashes illuminate the objects in absolute space at the corresponding instants of time just as strobe lighting illuminates dancers in a darkened room. In this computer-generated perspective view, the vertices of a triangle represent the positions of three mass points as they move through absolute space. The triangles formed by the points at successive instants are shown.

such that if the motions of the bodies are known at some instant, the laws determine all the future movements. All the world's history can be determined from two snapshots taken in quick succession. (If you know where something is at two closely spaced instants, you can tell its speed and direction. Two such snapshots thus encode the future.)

Newton's picture is close to everyday experience. We do not see absolute space and time, but we do see something quite like them – the rigid Earth, which defines positions, and the Sun, whose motion is a kind of clock. Newton's revolution was the establishment of strict laws that hold in such a framework.

LAWS AND INITIAL CONDITIONS

These laws have a curious property. They determine motions only if certain *initial conditions* are combined with them. Newton believed that God 'set up' (created) the universe at some time in the past by placing objects in absolute space with definite motions; after that, the laws of motion took over. The statement that Newton's is a clockwork universe is a bit misleading. Clocks have one predetermined motion: the pendulum of the grandfather clock simply goes backwards and forwards. The Newtonian universe is much more remarkable, being capable of many motions. However, once an initial condition has been chosen, everything follows.

Thus, there are two disparate elements in the scientific account of the universe: eternal laws, and a freely specifiable initial condition. Einstein's relativity and major astronomical discoveries have merely added to this dual scheme the exciting novelty of a universe exploding into being about fifteen billion years ago. The initial condition was set at the Big Bang.

Some people question this dual scheme. Is it an immutable feature? Might we not find laws that stand alone, without initial conditions? These questions are particularly relevant because Newton's laws (and also Einstein's theories of relativity, which replaced them) have a property that seems quite at variance with the way we feel the universe works – that the past determines the future. We do not think that causality works from the future to the past. Scientists always consider initial conditions. But Newton's and Einstein's laws work equally well in both directions. The truth is that the string of triangles in Figure 1 is determined by Newton's laws acting in both directions by any two neighbouring triangles anywhere along the string. You can persuade yourself of this by looking at the figure again. It is impossible to say in which direction time flows. The caption speaks of 'strobe lighting' illuminating the triangles at equal time intervals, but does not say which is illuminated first. Scientists could examine the triangles until the crack of doom but could never find which came first. This is related to one of the biggest puzzles in science.

WHY IS THE UNIVERSE SO SPECIAL?

The universe we see around us today is special: it is very highly ordered. For example, light streams away in a very regular flow from billions upon billions of stars throughout the universe. These stars are themselves collected together in galaxies, of which there are just a few basic types. Here on Earth we find very complex molecules and very complicated life forms that could not possibly exist were it not for the steady stream of sunlight that constantly bathes our planet. However, the vast majority of conceivable initial conditions there could have been at the Big Bang would have led to universes much less interesting – indeed, positively dull – compared with ours. Only an exceptional initial condition could have led to the present order. That is the puzzle. Modern science is in the remarkable position of possessing beautiful and very well tested laws without really being able to explain the universe. In the dual scheme of laws and initial conditions, the great burden of explaining why the universe is as it is falls to the initial conditions. Science can as yet give no explanation of why those conditions were as they must have been to explain the presently observed universe. The universe looks like a fluke.

There are two remarkable things about the order in the universe: the amount of it and the way it degrades. One of the greatest discoveries of science, made about a hundred and fifty years ago, was the second law of thermodynamics. Studies of the efficiency with which steam engines turn heat into mechanically useful motion led to the concept of *entropy*. As originally discovered, this is a measure of how much useful work can be got out of hot gas, say. It is here that the arrow of time, which we know from direct experience, enters physics. Almost all processes observed in the universe have a directionality. In an isolated system, temperature differences are always equalized. This means, for example, that you cannot extract energy from a cooler gas to make a hotter gas even hotter and chuff along in your steam engine even faster. More strictly, if you did, you would degrade more energy than you gain and finish up worse off.

I have already mentioned the unidirectional process of a cup breaking. Another is mixing cream with coffee. It is virtually impossible to reverse these processes. This is beautifully illustrated by running a film backwards: you see things that are impossible in the real world. This unidirectionality, or arrow, is precisely reflected in the fact that the entropy of any isolated system left to itself always increases (or perhaps stays constant).

It was recognized in the late nineteenth century that this unidirectionality of observed processes was in sharp conflict with the fact that Newton's laws should work equally well in either time direction. Why do natural processes always run one way, while the laws of physics say they could run equally well either way? For four decades, from 1866 until his suicide on 5 September 1906 in the picturesque Adriatic resort of Duino, the Austrian physicist Ludwig Boltzmann attempted to resolve this conflict. He introduced a theoretical definition of entropy as the *probability* of a state. He firmly believed in atoms – the existence of which remained controversial until the early years of the twentieth century – conceived of as tiny particles rushing around at great speed in accordance with Newtonian laws. Heat was assumed to be a measure of the speed of atoms: the faster the atoms, the hotter the substance. By the second half of the nineteenth century, physicists had a good idea of the immense number of atoms (assuming that they existed) there must be even in a grain of sand, and Boltzmann, among others, saw that statistical arguments must be used to describe how atoms behave.

He asked how probable a state should be. Imagine a grid of 100 holes into which you drop 1000 marbles at random. It is hugely improbable that they will all finish up in one hole. I am not going to give numbers, but it is simple to work out the probability that all will land in one hole or, say, in four adjacent holes. In fact, one can list every possible distribution of the marbles in the grid, and then see in how many of these distributions all the marbles fall in one hole, in four adjacent holes, eight adjacent holes, and so on. If each distribution is assumed to be equally probable, the number of ways a particular outcome can happen becomes the relative probability of that outcome, or state. Boltzmann had the inspired idea that, applied to atoms, this probability (which must also take into account the velocities of the atoms) is a measure of the entropy that had been found through study of the thermodynamics of steam engines.

There is no need to worry about the technical details. The important thing is that states with low entropy are inherently improbable. Boltzmann's idea was brilliantly successful, and much of modern chemistry, for example, would be unthinkable without it. However, his attempt to explain the more fundamental issues associated with the unidirectionality of physical processes was only partly successful.

He wanted to show that, matching the behaviour of macroscopic entropy, his microscopic entropy would necessarily increase solely by virtue of Newton's laws. This seems plausible. If a large number of atoms are in some unlikely state, say all in a small region, so that they have a low entropy, it seems clear that they will pass to a more probable state with higher entropy. However, it was soon noted that there are exactly as many dynamically possible motions of the atoms that go from states of low probability to states of high probability as vice versa. This is a straight consequence of the fact that Newton's laws have the same form for the two directions of time. Newton's laws alone cannot explain the arrow of time.

Only two ways have ever been found to explain the arrow: either the universe was created in a highly unlikely special state, and its initial order has been 'degrading' ever since, or it has existed for ever, and at some time in the recent past it entered by chance an exceedingly improbable state of very low entropy, from which it is now emerging. The second possibility is entirely compatible with the laws of physics. For example, if a collection of atoms (which obey Newton's laws) is confined in a box and completely isolated, it will, over a sufficiently long period of time, visit (or rather come arbitrarily near) all the states that it can in principle ever reach, even those that are highly ordered and statistically very unlikely. However, the intervals of time between returns to states of very low entropy are stupendously long (vastly longer than the presently assumed age of the universe), and neither explanation is attractive.

The fact is that mechanical laws of motion allow an almost incomprehensibly large number of different possible situations. Interesting structure and order arise only in the tiniest fraction of them. Scientists feel they should not invoke miracles to explain the order we see, but that leaves only statistical arguments, which give bleak answers (only dull situations can be expected), or the so-called anthropic principle that if the world were not in a highly structured but extremely unlikely state, we should not exist and be here to observe it.

One of my reasons for writing this book is that timeless physics opens up new ways of thinking about structure and entropy. It may be easier to explain the arrow of time if there is no time!

CHAPTER 2

Time Capsules

THE PHYSICAL WORLD AND CONSCIOUSNESS

The discussion in Chapter 1 prompts the question of how our sense of the passage of time arises. Before we can begin to answer this, we have to think about another mystery – consciousness itself. How does brute inanimate matter become conscious, or rather self-conscious?

No one has any idea. Consciousness and matter are as different as chalk and cheese. Nothing in the material world gives a clue as to how parts of it (our brains) become conscious. However, there is increasing evidence that certain mental states and activities are correlated with certain physical states in different specific regions of the brain. This makes it natural to assume, as was done long ago, that there is psychophysical parallelism: conscious states somehow reflect physical states in the brain.

Put in its crudest form, a brain scientist who knew the state of our brain would know our conscious state at that instant. The brain state allows us to reconstruct the conscious state, just as musical notes on paper can be transformed by an orchestra into music we can hear. By the 'state' of a system, say a collection of atoms, scientists usually mean the positions of all its parts and the motions of those parts at some particular instant. It is widely assumed that conscious states, in which, after all, we are aware of motion directly, are at the least correlated with (correspond to) brain states that involve not only instantaneous positions but also motions and, more generally, change (associated with flow of electric currents or chemicals, for example). This is a natural assumption. Our awareness of motion and change is vivid and often exciting: think of watching gymnastics, or the 100-metre sprint final in the Olympic Games. We suppose that the impression of motion must be created by some motion or change in the brain.

However, if the physical processes in the brain are controlled by laws like Newton's, such an assumption runs up against the problem that they distinguish no direction of time. Figure 1, with its impossibility of saying in which direction time flows, makes this clear. It is no help to go from its three particles to billions of them. Observed effects should have a real cause. The chain from cause to effects may be quite long and take surprising forms, but a cause there must be. It is unsatisfactory to suppose that we have a direct awareness of an invisible flow of time. Our sense of the passage of time and, even more basically, of seeing motion and knowing its direction, ought to have a cause we can get our hands on.

The lack of time direction in the bare laws of motion led Boltzmann to a remarkable suggestion (quoted in the Notes). As we have seen, Newtonian systems can enter highly ordered phases. These are exceptionally rare periods separated by 'deserts' of monotony. Nevertheless, every now and then a system will enter one. Its entropy will go down, reach a minimum, and then start to increase.

We should not think of this happening in a definite direction of time. Instead, we should picture the states of the system strung out in a line, as in Figure 1, which we could 'walk along' in either direction. Every now and then, with immense stretches between them, we will come upon regions in which the entropy decreases and the order increases. Then the entropy will start to increase again. Someone 'walking' in the opposite direction would have the same experience. Now, such a line of states can represent the entire universe, including human beings. Since we are very complicated and exhibit much order, we can be present only in the exceptional regions of low entropy.

Boltzmann's suggestion, startling when first encountered, was that conscious beings could exist on either side of a point of lowest entropy, and that the beings on both sides would regard that point as being in their past. Time would seem to increase in both directions from it. In this view, time itself neither flows nor has a direction; it is at most a line. It is only the instantaneous configurations of matter, strung out like washing on the line, that very occasionally suggest that time has a direction associated with it. The direction is in the washing, not the line. What is more, depending on the position in the line, the 'arrow' will point in opposite directions.

This, then, gives a genuine cause for our awareness of motion and the passing of time. The conscious mind, in any instant, is actually aware of a short segment of the 'line of time', along which there is an entropy gradient. Time seems to flow in the direction of increasing entropy. Interestingly, consciousness and understanding are always tied to a short time span, which was called the specious present by the philosopher and psychologist William James (brother of novelist Henry). The specious present is closely related to the phenomenon of short-term memory and our ability to grasp and understand sentences, lines of poems and snatches of melody. It has a duration of up to about three seconds.

The key element in Boltzmann's idea is comparison of structures. There needs to be qualitative change in the brain patterns along a segment of the 'line of time'. If the brain pattern in each instant is likened to a card, then the patterns become a pack of cards, and our conscious experience of time flow arises (somehow) from the change of pattern across the pack. Though we may not understand the mechanism, the effect does have a cause.

To summarize: Newtonian time is an abstract line with direction – from past to future. Boltzmann keeps the line but not the direction. That belongs to the 'washing'. But do we need the line?

TIME WITHOUT TIME

Perhaps not. The brain often fools us. When we first look at certain drawings, they appear to represent one thing. After a while, the image flickers and we see something different. The reason is well understood: the brain processes information before we get it. We do not see things as they are but as the brain interprets them for us. There are very understandable reasons for this, but the fact remains that we are often fooled by such 'deceptions'.

Could all motion be a similar deception? Suppose we could freeze the atoms in our brains at some instant. We might be watching gymnastics. What would brain specialists find in the frozen pattern of the atoms? They will surely find that the pattern encodes the positions of the gymnasts at that instant. But it may also encode the positions of the gymnasts at preceding instants. Indeed, it is virtually certain that it will, because the brain cannot process data instantaneously, and it is known that the processing involves transmission of data backwards and forwards in the brain. Information about the positions of the gymnasts over a certain span of time is therefore present in the brain in any one instant.

I suggest that the brain in any instant always contains, as it were, several stills of a movie. They correspond to different positions of objects we think we see moving. The idea is that it is this collection of 'stills', all present in any one instant, that stands in psychophysical parallel with the motion we actually see. The brain 'plays the movie for us', rather as an orchestra plays the notes on the score. I am not going to attempt to elaborate on how this might be done; all I want to do is get the basic idea across. There are two parts to it. First, each instantaneous brain pattern contains information about several successive positions of the objects we see moving in the world. These successive positions need correspond only to a smallish fraction of a second. Second, the appearance of motion is created by the instantaneous brain pattern out of the simultaneous presence of several different 'images' of the gymnasts contained within it (Figure 2). This happens independently of the earlier and later brain states.

Figure 2 My explanation of how it might be possible to 'see' motion when none is there is illustrated in this chronophotograph of a sideways jump. My assumption is that the pattern of the atoms in our brain encodes, at any instant, about six or seven images of the gymnast. The standard 'temporal' explanation is that the gymnast passes through all these positions in a fraction of a second. My idea is that when we think we are seeing actual motion, the brain is interpreting all the simultaneously encoded images and, so to speak, playing them as a movie.

This proposal is not so very different from Boltzmann's idea that the sense of motion is created from several qualitatively different patterns arranged along the 'line of time'. Instead, I am suggesting that it is created by the brain from the juxtaposition of several subpatterns within one pattern. The arrow of time is not in the washing line, it is not in several pieces of washing, it is in each piece. If we could preserve one of these brain patterns in aspic, it would be perpetually conscious of seeing the gymnasts in motion. If you find this idea a bit startling, I am glad because I find it does bring home the 'freezing of motion' that I think we have to contemplate. In fact, since brain function and consciousness are fields in which I have no expertise, I would like you to regard this suggestion in the first place as a means of getting across an idea, the main application of which I see in physics.

To that end, I want to introduce the notion of special Nows, or *time capsules*, as I call them.

TIME CAPSULES

By a time capsule, I mean any fixed pattern that creates or encodes the appearance of motion, change or history. It is easiest to explain the idea by examples, for example the *Ariel* in the storm in Turner's painting. Although they are all static in themselves, pictures often suggest that something has happened or is happening – with a vengeance in this painting. But in reality it simply is. I know no better example of something static that gives the impression of motion.

In pictures, the impression is deliberately created. Much more significant for my purposes are time capsules that arise naturally and have to be interpreted, by the examination of records they seem to contain. Records, or apparent records, play a vital role in my idea that time is an illusion. I use *records* primarily in the sense of, for example, fossils, which occur naturally and are interpreted by us as relics of things that actually existed. Less directly, all geological formations, rock strata in particular, are now invariably interpreted by geologists as constituting a record (to be interpreted) of past geological processes. Finally, there are records that people create deliberately: doctors' notes, minutes of committee meetings, astronomical observations, photographs, descriptions of the initial and final conditions of controlled experiments, and so on. All such things, and many more, I call records. My position is that the things we call records are real enough, and so is their structure. They are the genuine cause of our belief in time. Our only mistake is the interpretation: time capsules have a cause, but time is no part of it.

Let me now attempt a more formal definition. Any static configuration that appears to contain mutually consistent records of processes that took place in a past in accordance with certain laws may be called a time capsule. From my point of view, it is unfortunate that the dictionary definition (in *Webster's*) of a time capsule is 'a container holding historical records or objects representative of current culture that is deposited (as in a cornerstone) for preservation until discovery by some future age'. I do not mean that. But we have all had the experience of walking into a house untouched by historical development for decades or centuries and declaring it to be a perfect time capsule. This, I believe, happens to us in each instant of time we experience. The only difference is that we experience our current time capsule, not someone else's. And we are mistaken in the way we interpret the experience.

It is important for me that, as I point out in the next section, the phenomenon of time capsules is very widespread in the physical world, and is not restricted to our mental states and experiences. In addition to my caveat at the end of the previous section, I should emphasize that I am not claiming consciousness plays some remarkable novel or extraphysical role in the world. Unlike Roger Penrose in his best-seller *The Emperor's New Mind*, I am not suggesting that there is any 'new physics' associated with mental states. There may be, but that is not part of my time-capsule idea. However, I do believe we have to think carefully about the role of consciousness in the picture that we form of the world.

First, all knowledge and theorizing comes to us through the conscious state. If we want to form an overall picture of things, we cannot avoid allotting a place to consciousness. It is necessary for completeness: we have to consider 'where we stand'. This is closely related to a second factor. Viewed as a physical system, the brain is organized to an extraordinary degree. It is vastly more complicated and intricate than the air we breathe or the star clusters we see through telescopes. There may not be any locations anywhere in the universe that are more subtly and delicately organized than human brains. There is not merely the brain structure as such, but also the distillation of accumulated human experience and culture that we carry in our brains. But this very organization may be giving us a distorted picture of the world. If you stand, like Turner bound to the *Ariel*'s mast, in the tornado's maelstrom, you might well suspect that the universe is just one great whirlpool.

The lesson we learned from Copernicus, Kepler and Galileo is here very relevant. They persuaded us, against what seemed to be overwhelming evidence to the contrary, that the Earth moves. They taught us to see motion where none appears. The notion of time capsules may help us to reverse that process – to see perfect stillness as the reality behind the turbulence we experience.

Stand, as I have with a daughter, and look at Jupiter against the winter stars. Every clear frosty night we stood on the utterly motionless Earth – as it appeared to our senses – and watched through the winter as Jupiter, high in the sky, tracked night by night eastwards against the background of the stars. But then Jupiter slowed down, came to a stop, and went backwards in the retrograde motion that so puzzled the ancients. Then this motion stopped, and the eastwards motion recommenced. In all this Jupiter moved, not us. We could see it with our eyes. Seeing is believing. But what did Copernicus say? We must be careful not to attribute to the heavens (Jupiter) what is truly in the Earth-bound observer. I could persuade my daughter that the motion of the Earth, not of Jupiter, gives rise to the retrograde motion. To interpret events, we must know where we stand and understand how that affects what we witness. But we observe the universe from the middle of a most intricate processing device, the human brain. How does that affect our interpretation of what we see?

EXAMPLES OF TIME CAPSULES

As a first example, we can stay within the brain but consider long-term memory. A game we sometimes play at Christmas brings out the importance of mutually consistent records held in structures. Fifty events in recent world history are written down on separate cards without dates attached. Players are divided into teams and given the cards jumbled up. The challenge is to put them in the correct chronological order. The only resource each team has to attack the task is their collective long-term memories, which every good realist (myself included) will surely agree are somehow or other 'hard-wired' into their brains. How each team fares depends on the consistency of its members' recollections – the records in their brains.

This example shows clearly that all we know about the past is actually contained in present records. The past becomes more real and palpable, the greater the consistency of the records. But what is the past? Strictly, it is never anything more than we can infer from present records. The word 'record' prejudges the issue. If we came to suspect that the past is a conjecture, we might replace 'records' by some more neutral expression like 'structures that seem to tell a consistent story'.

The relevance of this remark is brought home by the sad examples of brain damage that takes away the ability to form new memories but leaves the existing long-term memory intact. One patient, still alive, retains good memory and a sense of himself as he was before an operation forty years ago, but the rest is blank. It is possible to have meaningful discussions about what are for him current events even though they are all those years away, but the next day he has no recollection of the discussion. The mature brain is a time capsule. History resides in its structure.

After our own brains, the most beautiful example of a time capsule that we know intimately is the Earth – the whole Earth. Above all, I am thinking of the geological and fossil records in all their multifarious forms. What an incredible richness of structure is there, and how amazingly consistent is the story it tells. I find it suggestive that it was the geologists – not the astronomers or physicists – who first started to suggest an enormous age for the Earth. They were the discoverers of deep time, which did start as conjecture. And it was all read off from rocks, most of which are still with us now, virtually unchanged from the form they had when the geologists reached those conclusions. The story of the antiquity of the Earth and of its creation from supernova debris – the stardust from which we believe we ourselves are made – is a story of patient inference built upon patient inference based upon marks and structures in rocks. On this rock – the Earth in all its glory – the geologists have built the history of the world, the universe even.

What is especially striking about the Earth is the way in which it contains time capsules nested within time capsules, like a Russian doll. Individual biological cells (properly interpreted) are time capsules from which biologists read genetic time. Organs within the body are again time capsules, and contain traces of the history and morphogenesis of our bodies. The body itself is a time capsule. History is written in a face, which carries a date – the approximate date of our birth. We can all tell the rough age of a person from a glance at their face. Wherever we look, we find mutually consistent time capsules – in grains of sand, in ripe cherries, in books in libraries. This consistent meshing of stories even extends far from the Earth and into the outermost reaches of the universe. The abundances of the chemical elements and isotopes in the gas of stars and the waters of the oceans tell the story of the stars and a Big Bang that created the lightest elements. It all fits together so well.

For me, two facts above all stand out from this miracle of nature. If we discount the direct perception of motion in consciousness, all this fantastic abundance of evidence for time and history is coded in static configurational form, in structures that persist. This is the first fact, and it is ironic. The evidence for time is literally written in rocks. This is why I believe the secret of time is to be unravelled through the notion of time capsules. It is also the reason why I seek to reduce the other hard and persistent evidence for time and motion – our direct awareness of them in consciousness – to a time-capsule structure in our brains. If I can make such a structure responsible for our short-term memory – the phenomenon of the specious present – and for the actual seeing of motion, then all appearances of time will have been reduced to a common basis: special structure in individual Nows.

The second fact that needs to be taken on board is the sheer creativity of Nature. How does Nature create this rich, rich structure that speaks to us so insistently and consistently of time? How could it and we come to be if there is no time? The appearance of time is a deep reality, even without the motion we see and the passage of time we sense in consciousness. It is written all over the rocks. Any plausible account of the universe must, first and foremost, explain the existence of the structures we see and the semantic freight (i.e. the seemingly meaningful story) that they carry.

If we can explain how they arise, time capsules offer the prospect of a much more radical explanation of the properties of time than Boltzmann's account of the origin of its arrow. To explain the appearance of an arrow, he still had to assume a succession of instants strung out along a 'line of time'. I have already suggested that the line may be redundant. The inference that it exists can emerge from a single Now. The instant is not in time – time is in the instant.

CHAPTER 3

A Timeless World

FIRST OUTLINE

Now I want to start on the attempt to show you that, at least as a logical possibility, the appearance of time can arise from utter timelessness. I shall do this by comparing two imaginary exercises. I begin by presenting you with two bags, labelled Current Theory and Timeless Theory. When you open them up, you find that each bag is filled with cardboard triangles, all jumbled up. Now, triangles come in all shapes and sizes. The first thing you notice is that the first bag contains far fewer triangles than the second. Closer examination reveals that the two collections are very different. Let me begin by describing the contents of Current Theory.

First, you notice that it contains triangles of all different sizes. There is a smallest triangle, very tiny; then another very like it, but a little larger and with a slightly different shape; and so on. In fact, you soon realize that you can lay out all the triangles in a sequence. The order in which they should go is clear because each successive triangle differs only slightly from its predecessor. Their increasing size makes the ordering especially easy. Of course, a real bag can contain only finitely many triangles, but I shall suppose that there are infinitely many and that the sequence is endless, the triangles getting ever larger.

Such a sequence of triangles is like the sequence of experienced instants that I suggested 'photographing'. It is also like the succession of Newtonian instants from the moment God decided to create the universe, or the succession of states of the universe expanding out of the Big Bang, represented by the smallest triangle. In fact, the contents of Current Theory correspond to the simplest Newtonian universe that can begin to model the complexity of the actual universe: three mass points moving in absolute space and time, as in Figure 1. Initially very close to each other, they move apart so rapidly that gravity cannot pull them back, and they fly off to infinity.

According to Newton, the three mass points are, at all instants, at certain positions in absolute space and form certain triangles. The triangles tell us how the points are placed relative to one another, but not where they are in absolute space. It is such triangles, represented in cardboard, that I imagine have been put into the Current Theory bag. Since we cannot experience absolute space and time directly, I have tried to match the model more closely to our actual experience. The sequence of triangles corresponds to one possible history. There could be many such histories that match the dual scheme of laws and initial conditions. But we find only one in the Current Theory bag.

Next, we examine the Timeless Theory bag. There are two big differences. First, it contains vastly more triangles (it could, in fact, contain all conceivable triangles). More significantly, there are so many of them that it is quite impossible to arrange them in a continuous sequence. Second, the triangles are present in multiple copies. That is, we might, after a very extensive search, find ten identical copies of one particular triangle, two of another, and ten million of yet another. That is really the complete story. It is all that most people would notice.

I think you will agree that the Current Theory bag does match experience quite closely. The triangles stand for each of the instants you experience, and they follow one another continuously, just as the instants do. By giving them to you in a bag and getting you to lay them out in a sequence, I am giving you a 'God's eye' view of history. All its instants are, as it were, spread out in eternity as if you surveyed them from a mountain-top. In fact, this way of thinking about time has long been a commonplace among Christian theologians and some philosophers, and has prompted them to claim that time does not exist but that its instants all exist together and at once in eternity. My claim is much stronger. I am saying that reality, if we could see all of it, is not at all like the contents of the Current Theory bag with its single sequence of states. It is like the contents of the Timeless Theory bag, in which in principle all conceivable states can be present. Nothing in it resembles our experience of history as a unique sequence of states: that experience is usually explained by assuming that there is a unique sequence of states. I deny that there is such a sequence, and propose a different explanation for the experience that prompts us to believe in it. The only thing the bags have in common with our direct experience of time is the parallel between individual triangles as models of individual instants of time.

Actually, the bags share another property – their contents satisfy a law. Given the sequence of triangles of the first bag, clever mathematicians could deduce that they correspond to the triangles formed by three gravitationally interacting bodies. They could even reconstruct the bodies' positions in absolute space, and the amount of time that elapses between any two of the triangles in the sequence. With the second bag, mathematicians would discover that the numbers in which the different triangles occur are not random – chosen by chance – but satisfy a law. The numbers vary from triangle to triangle in an ordered fashion. But at first glance at least, this law seems to have no connection with the law that creates the unique sequence of triangles in the first bag. Also, there is nothing like the dual scheme of law and initial condition that creates the sequence of the first bag. In a sense that I shall not yet try to explain, there is just a law, with nothing like an initial condition that has to be added to it.

How is the appearance of time ever going to emerge from the contents of the Timeless Theory bag as just described? Bare triangles lying in a jumbled heap certainly cannot make that miracle happen. Triangles have a structure that is much too simple. This is why I said that rich structure ordered in a special way is an essential element if a notion of time is to emerge. If, when we open the Timeless Theory bag, we find it contains, not triangles, but vastly richer structures, some of which are time capsules in the sense I have defined, my task does not seem quite so hopeless. By definition, time capsules suggest time. But finding just a few time capsules in a vast heap of otherwise nondescript structures will not get me very far.

This is where the assumption that all the structures found in the bag come in multiple copies, and that the numbers of these copies, which can vary very widely, are determined by a definite timeless rule, becomes crucial. Imagine that all the structures for which the numbers of identical copies in the bag are large are time capsules, while there are few copies of structures that are not time capsules. Since the overwhelming majority of possible structures that can exist are certainly not time capsules, any rule that does fill the bag with time capsules will be remarkably selective, creative one might say. If, in addition, you can find evidence that the universe is governed by a timeless law whose effect is to discriminate between structures and which actually selects time capsules with surprising accuracy, then you might begin to take such ideas more seriously. You might begin to see a way in which the Timeless Theory could still explain our experience of time, and could perhaps be superior to the Current Theory.

However, you will probably dismiss such a possibility as the wildest fantasy. Why should Nature go to such contrived lengths simply to create an impression of time and fool poor mortals? To counter this natural reaction, let me give a little more detail about those hints of the non-existence of time that I mentioned in Chapter 1. This may at least persuade you that some dramatic change could be in the offing.

THE CRISIS OF TIME

Physics is regarded as the most fundamental science. It is an attempt to create a picture of reality as we should see it if we could, somehow, step out of ourselves. For this reason it is rather abstract. In addition, it often deals with conditions far removed from everyday human experience – deep inside the atom, where quantum theory holds sway, and in the far-flung reaches of space, where Einstein's general relativity reigns. The ideas I want to tell you about have come from attempts during the last forty years to unite these two realms (Box 2). They have produced a crisis. The very working of the universe is at stake: it does not seem to be possible, in any natural and convincing way, to give a common description of them in which anything like time occurs.

Frustratingly little progress has been made. However, in 1967 a possible picture did emerge from a paper by the American Bryce DeWitt. He found an equation that, if his reasoning is sound, describes the whole universe – both atoms and galaxies – in a unified manner. Because John Wheeler, the American physicist who coined the term 'black hole', played a major part in its discovery, this equation is called the Wheeler–DeWitt equation. It is controversial in at least three respects. First, many experts believe that the very derivation of the equation is flawed – that it was obtained by an invalid procedure. Second, the equation is not yet even properly defined, as there are still many technical difficulties to be overcome. In fact, it is more properly regarded as a conjecture: a tentative proposal for an equation that is not yet proved. And third, the experts argue interminably over what meaning it might have and whether it can ever be promoted to the status of a bona fide equation. Ironically, DeWitt himself thinks that it is probably not the right way to go about things, and he generally refers to it as 'that damned equation'. Many physicists feel that a different route, through so-called superstring theory, which it is hoped will establish a deep unity between all the forces of nature, is the correct way forward. That many of the best physicists have concentrated on superstring theory is probably the main reason why the 'crisis of time' brought to light by the Wheeler–DeWitt equation has not attracted more attention. However, there is no doubt that the equation reflects and unifies deep properties of both quantum theory and general relativity. Quite a sizeable minority of experts take the equation seriously. In particular, much of the work done by Stephen Hawking in the last twenty years or so has been based on it, though he has his own special approach to the problem of time that it raises.

For now, all I want to say about the Wheeler–DeWitt equation is that if one takes it seriously and looks for its simplest interpretation, the picture of the universe that emerges is like the contents of the Timeless Theory bag. For a long time, physicists shied away in distrust from its apparently timeless nature, but during the last fifteen years or so a small but growing number of physicists, myself included, have begun to entertain the idea that time truly does not exist. This also applies to motion: the suggestion is that it too is pure illusion. If we could see the universe as it is, we should see that it is static. Nothing moves, nothing changes. These are large claims, and the bulk of my book will discuss the arguments from physics (presented as simply as I can) that lead me and others to such conclusions. At the end, I shall outline, through the notion of time capsules, a theory of how a static universe can nevertheless appear to teem with motion and change.

Now I want to give you a better feel for what a timeless universe could be like. What we need first is a proper way to think about Nows.

THE ULTIMATE ARENA

One issue that runs through this book is this: what is the ultimate arena of the universe? Is it formed by space and time (space-time), or something else?

This is the issue raised by Dirac's sentence I quoted in the Preface: 'This result has led me to doubt how fundamental the four-dimensional requirement in physics is.' I believe that the ultimate arena is not space-time. I can already begin to give you an idea of what might come in its place.

I illustrated the Newtonian scheme by a model universe of just three particles. Its arena is absolute space and time. The Newtonian way of thinking concentrates on the individual particles: what counts are their positions in space and time. However, Newton's space and time are invisible. Could we do without them? If so, what can we put in their place? An obvious possibility is just to consider the triangles formed by the three particles, each triangle representing one possible relative arrangement of the particles. These are the models of Nows I asked you to contemplate earlier. We can model the totality of Nows for this universe by the totality of triangles. It will be very helpful to start thinking about this totality of triangles, which is actually an infinite collection, as if it were a country, or a landscape.

If you go to any point in a real landscape, you get a view. Except for special and artificial landscapes, the view is different from each point. If you wanted to meet someone, you could give them a snapshot taken from your preferred meeting point. Your friend could then identify it. Thus, points in a real country can be identified by pictures. In a somewhat similar way, I should like you to imagine Triangle Land. Each point in Triangle Land stands for a triangle, which is a real thing you can see or imagine. However, whereas you view a landscape by standing at a point and looking around you, Triangle Land is more like a surface that seems featureless until you touch a point on it. When you do this, a picture lights up on a screen in front of you. Each point you touch gives a different picture. In Triangle Land, which is actually three-dimensional, the pictures you see are triangles. A convenient way of representing Triangle Land is portrayed in Figures 3 and 4.

I have gone to some trouble to describe Triangle Land because it can be used to model the totality of possible Nows. Like real countries, and unlike absolute space, which extends to infinity in all directions, it has frontiers. There are the sheets, ribs and apex of Figure 4. They are there by logical necessity. If Nows were as simple as triangles, the pyramid in Figure 4 could be seen as a model of eternity, for one notion of eternity is surely that it is simply all the Nows that can be, laid out before us so that we can survey them all.

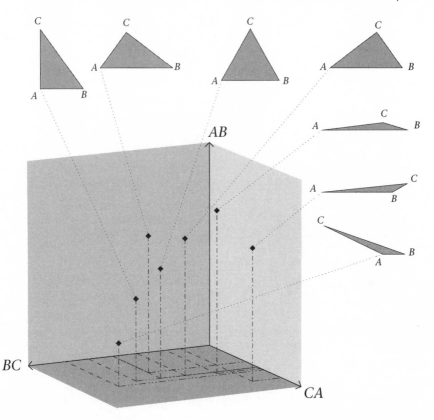

Figure 3 The seven triangles represent several possible arrangements of a model universe of three particles *A*, *B*, *C*. Each triangle is a possible Now. Each Now is associated with a point (black diamond) in the 'room' formed by the three grid axes *AB*, *BC*, *CA*, which meet at the corner of the 'room' farthest from you. The black diamond that represents a given triangle *ABC* is situated where the distance to the 'floor' is the length of the side *AB* (measured along the vertical axis), and the distances to the two 'walls' are equal to the other two sides, *BC* and *CA*. The dash-dotted lines show the grid coordinates. In this way, each model Now is associated with a unique point in the 'room'. As explained in the text, if you 'touched' one of the black diamonds, the corresponding triangle would light up. However, not every point in the 'room' corresponds to a possible triangle – see Figure 4.

A three-particle model universe is, of course, unrealistic, but it conveys the idea. In a universe of four particles, the Nows are tetrahedrons. Whatever the number of particles, they form some structure, a *configuration*. Plastic balls joined by struts to form a rigid structure are often used to model molecules, including macromolecules such as DNA, which are 'megamolecules'. You can move such a structure around without

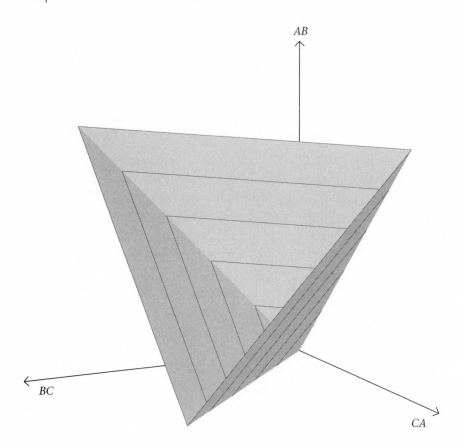

Figure 4 This shows the same 'room' and axes as in Figure 3, but without the walls shaded. Something more important is illustrated here. In any triangle, no one side can be longer than the sum of the other two. Therefore, points in the 'room' in Figure 3 for which one coordinate is larger than the sum of the other two do not correspond to possible triangles. All triangles must have coordinates inside the 'sheets' spanned between the three 'ribs' that run (towards you) at 45° between the three pairs of axes *AB*, *BC* (up to the left), *AB*, *CA* (up to the right) and *BC*, *CA* (along the 'floor', almost towards you). Points outside the sheets do not correspond to possible triangles. However, points on the sheets, the ribs and the apex of the pyramid formed by them correspond to special triangles. If vertex *A* in the thin triangle at the bottom right of Figure 3 is moved until it lies on *BC*, the triangle becomes a line, which is still just a triangle, because *BC* is now equal to (but not greater than) the sum of *CA* and *AB*. Such a triangle is represented by a point on one of the 'sheets' in Figure 4. If point *A* is then moved, say, towards *B*, the point representing the corresponding triangle in Figure 4 moves along the 'sheet' to the corresponding 'rib', which represents the even more special 'triangles' for which two points coincide. Finally, the apex, where the three ribs meet in the far corner of the 'room', corresponds to the unique and most special case in which all three particles coincide. Thus, Triangle Land has a 'shape' which arises from the rules that triangles must satisfy. The unique point at which the three particles coincide I call *Alpha*.

changing its shape. For any chosen number of balls, many different structures can be formed. That is how I should like you to think about the instants of time. Each Now is a structure.

For each definite collection of structures – triangles, tetrahedrons, molecules, megamolecules – there is a corresponding 'country' whose points correspond to them. The points are the possible configurations. Each configuration is a possible thing; it is also a possible Now. Unfortunately it is impossible to form any sort of picture of even Tetrahedron Land: unlike Triangle Land, which has three dimensions, it has six dimensions. For megamolecules, one needs a huge number of dimensions. In Tetrahedron Land you could 'move about' in its six dimensions. As in my earlier example, the way to think about its individual points is that if you were to touch any one of them, a picture of the tetrahedron to which it corresponds would 'light up'. In any Megamolecule Land, with its vast number of dimensions, 'touching a point' would cause the corresponding megamolecule to 'light up'. The more complicated the structures, the greater the number of dimensions of the 'land' that represents them. However, the structures that 'light up' are themselves always three-dimensional.

You do not need to try to imagine these much larger spaces – Triangle Land will do. I hope you do not find it a dull structure or too hard to grasp. It is, in fact, an example of a very basic notion in physics called a *configuration space* that is normally regarded as too abstract to attempt to explain in books for non-scientists. But I cannot begin to get across to you my vision of a timeless universe without this concept. If you can get your mind round this concept – and I do encourage you to try – you will certainly understand a lot of my book. The notion of configuration space opens up a wonderfully clear way to picture, all at once, everything that can possibly be.

It will also give us new notions of time and history, stripping away and revealing as redundant the Newtonian superstructure. The observable history of a three-particle universe, when the invisible absolute space and time are abstracted away, is just a continuous sequence of triangles. Suppose we are given such a history. We can then mark, or plot, the points in Triangle Land that correspond to the triangles. We shall obtain a curve that winds around within the pyramid in Figure 4. In this new picture, history is not something that happens in time but a path through a landscape. A path is just a continuous track of points in a land. In this book I use the word *path* very often in the generalized sense of a continuous series of configurations taken by some system (consisting, usually, of material points). Understood in this sense, paths are possible histories. There is no time in this picture.

Paths highlight the dilemma brought to light by Boltzmann's work. On any path, you can call the point where you stand Now. But you can walk along a path in either direction. There is nothing in the notion of a path that can somehow make it a one-way street. You can also see that the notion of a moving present may be redundant. You might try to represent it by a spot of light moving along the path, making each successive point on the path into the present Now, and therefore more real than the 'past', through which the spot has already passed, and the 'future', which the spot has not yet reached. But if, as I have suggested, all our conscious experiences have their origin in real structure within the Nows, we can do without the fiction of the moving present. The sense we have that time has advanced to the present Now is simply our awareness of being in that Now. Different Nows give rise to different experiences, and hence to the impression that the time in them is different.

I need a name for the land of Nows. Plato, who lived about a century after Heraclitus and Parmenides, taught that the only real things are *forms* or *ideas*: perfect paradigms, existing in a timeless realm. In our mortal existence we catch only fleeting glimpses of these ideal forms. Now each point – each thing – in these 'countries' I have asked you to imagine could be regarded as a Platonic form. Triangles certainly are. I shall call the corresponding 'country' *Platonia*. The name reflects its mathematical perfection and timeless landscape. Nothing changes in Platonia. Its points are all the instants of time, all the Nows; they are simply there, given once and for all.

Platonia is vast. Size alone is insufficient to convey its vastness. Triangle Land already has three dimensions, and stretches out to infinity from its apex and frontiers. That reflects the already huge number of ways in which three objects can be arranged in space. As the number of objects is increased, the number of ways in which they can be arranged increases incredibly fast. The numbers one encounters in astronomy are as nothing compared with the number of possible arrangements of large numbers of objects. The instants of time are numberless. And each is different.

There is a saying about time, apparently first expressed in a piece of graffiti and much loved by John Wheeler, that seems apt here: 'Time is nature's way of preventing everything from happening all at once.' In a timeless world, verbs of becoming like 'happen' have no place. But if Nows are both concrete and distinct, it is a logical contradiction to suppose that they could 'happen at once', i.e. be superimposed on one another. I believe that the aphorism expresses a profound truth.

Developing the 'Platonic' theme, I conjecture that the actual universe in which we find ourselves corresponds to some Platonia. We have not yet fully grasped the structure of its points, its Nows. Perhaps we never shall, but I assume that in any instant what we experience, including the appearance of motion, is a transmuted representation of a part of one such Now. This is not far removed from Plato's original idea that we mortals are like beings confined from birth to a cave, and that all we ever comprehend of the outside world and the real beings in it are the shadows they cast on the wall of our cave as they pass its entrance. I also think that Plato was right when he said that Being (one of his forms, one of my instants of time) is real, but that Becoming is an illusion. However, I go further than Plato in attributing the illusion of Becoming to something that is real – a special time-capsule structure of Nows. The illusion of Becoming has its basis in real structure in special Being.

Platonia is the arena that I think must replace space and time. Why this should be so, how it can be done, and what physics in Platonia is like is the meat of the book. But it is already possible to see how differently creation and a supposed beginning of time appear in Platonia. Most people are baffled that time could begin. How many times do we hear the question, 'But what happened before the Big Bang?' The question reveals the depth to which the notion of an eternally flowing time is ingrained in the psyche. This is why I call the instants of time 'things', so as to break the spell, and why I have chosen the name Platonia for our home. It is also why I use paths as the image of history. In itself, there are no paths in Platonia, just as there were no paths on Earth before animals made them. The points of Platonia – the Nows – are worlds unto themselves. No thread of time joins them up. We must think of Newtonian-type dynamics as something that 'paints a path' onto the timeless landscape of Platonia.

Once the instinctive notion of time is expunged, it is easy to see that history, as a path in Platonia, can certainly start or end. The path to Land's End does terminate there: only the sea lies beyond. Triangle Land has a point like Land's End: it is the apex of the pyramid, which in

Figure 4 I called Alpha. Beyond it is nothing, not even sea. Looking for time before the Big Bang is like looking for Cornwall in the Irish Sea. If we think that time exists and increases or decreases along a path in Triangle Land that terminates at that apex, then we can see that time will certainly begin or end at that point. I think this is how we should think about the Big Bang. It is not in the past, it is at a kind of Land's End.

All Platonias seem by necessity to possess a distinguished point like the apex of Triangle Land. This is why I call it Alpha. It is suggestive that Platonia has an Alpha but no Omega: there is no limit to the size or complexity of things that can exist. Triangle Land opens out from Alpha to infinity, as do all Platonias. To underline this fact, Figure 5 is my own attempt to give a somewhat more artistic and simultaneously realistic representation of the actual Platonia of our universe, which of necessity is vastly more richly structured than Triangle Land.

Now we must begin to consider how the notion of Platonia will change the way we think about such seemingly simple things as motion. How can it emerge from a scheme without a vestige of time? Is motion really a pure illusion? If we were in London yesterday and New York today, we must have moved. Motion must exist. Let me persuade you that it does not.

IS MOTION REAL?

We had a cat called Lucy, who was a phenomenal hunter. She could catch swifts in flight, leaping two metres into the air. She was seen in the act twice, and must have caught other victims since several times we found just the outermost wing feathers of swifts by the back door. Faced with facts like this, isn't it ridiculous to claim there is no motion?

The argument seems decisive because we instinctively feel that Lucy has (or, rather had, since sadly she was killed by a car) some unchanging identity. But is the cat that leaps the cat that lands? Except for the changes in her body shape, we do not notice any difference. However, if we could look closely we might begin to have doubts. The number of atoms in even the tiniest thing we can see is huge, and they are in a constant state of flux. Because large numbers play a vital role in my arguments, I shall give two illustrations. Have you ever tried to form a picture of the number of atoms in a pea?

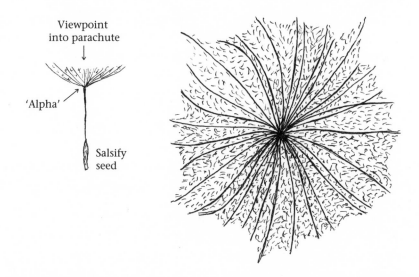

Figure 5 Triangle Land is like an inverted pyramid, with frontiers formed by special triangles as explained in Figure 4. Platonias corresponding to configurations of more than three particles have not only frontiers but also analogous internal topographic features. This illustration, based on the parachute of a salsify seed (shown life-size on left) from my wife's garden, is an attempt to give some idea of the rich structure of the frontiers of Platonia. No attempt is made to represent the even richer internal structure. Platonia's Alpha is where the ribs converge. Because Platonia has no Omega, the salsify ribs should extend out from Alpha for ever. (The wind carries the actual seeds rather efficiently into our neighbours' gardens, where the progeny flourish, but they are not always welcome, although salsify is an excellent vegetable.)

Imagine a row of dots a millimetre apart and a metre long. That will be one thousand dots (10^3). (Actually, it will be 1001, but let us forget the last 1.) One thousand such rows next to one another, also a millimetre apart, gives a square metre of dots, one million (10^6) in total. The number of dots in one or two squares like that is about the number of pounds or dollars ordinary mortals like me can hope to earn in a lifetime. Now stack one thousand such squares into a cube a metre high. That is already a billion (10^9). So it is surprisingly easy to visualize a billion. Five such cubes are about the world's human population. Yet we are nowhere remotely near the number of atoms in a pea.

We shall keep trying. We make another cube of these cubes. One thousand of them stretched out a kilometre long takes us up to a trillion (10^{12}). A square kilometre of them will be 10^{15} (about the number of cells in the human body), and if we pile them a kilometre high we get to 10^{18}.

We still have a long way to go. Make another row of one thousand of these kilometre cubes, and we get to 10^{21}. Finally, make that into a square, one thousand kilometres by one thousand kilometres and a kilometre high – it would comfortably cover the entire British Isles to that height. At last we are there: the number of dots we now have (10^{24}) is around the number of atoms in a pea. To get the number in a child's body, we should have to go up to a cube a thousand kilometres high. It hardly bears thinking about.

Equally remarkable is the order and organized activity in our bodies. Consider this extract from Richard Dawkins's *The Selfish Gene*:

> The haemoglobin of our blood is a typical protein molecule. It is built up from chains of smaller molecules, amino acids, each containing a few dozen atoms arranged in a precise pattern. In the haemoglobin molecule there are 574 amino acid molecules. These are arranged in four chains, which twist around each other to form a globular three-dimensional structure of bewildering complexity. A model of a haemoglobin molecule looks rather like a dense thornbush. But unlike a real thornbush it is not a haphazard approximate pattern but a definite invariant structure, identically repeated, with not a twig nor a twist out of place, over six thousand million million million times in an average human body. The precise thornbush shape of a protein molecule such as haemoglobin is stable in the sense that two chains consisting of the same sequences of amino acids will tend, like two springs, to come to rest in exactly the same three-dimensional coiled pattern. Haemoglobin thornbushes are springing into their 'preferred' shape in your body at a rate of about four hundred million million per second and others are being destroyed at the same rate.

If, as I think they must be, things are properly considered in Platonia, Lucy never did leap to catch the swifts. The fact is, there never was one cat Lucy – there were (or rather are, since Lucy is in Platonia for eternity, as we all are) billions upon billions upon billions of Lucys. This is already true for the Lucys in one leap and descent. Microscopically, her 10^{26} atoms were rearranged to such an extent that only the stability of her gross features enables us to call her one cat. What is more, compared with her haemoglobin molecules the features by which we identified her – the sharp eyes, the sleek coat, the wicked claws – were gross. Because we do not and cannot look closely at these Lucys, we think they are one. And all these Lucys are themselves embedded in the vast individual Nows of the universe. Uncountable Nows in Platonia contain something we should call Lucy, all in perfect Platonic stillness. It is because we abstract and 'detach' one Lucy from her Nows that we think a cat leapt. Cats don't leap in Platonia. They just are.

You might argue that even if cats do not have a permanent identity, their atoms do. But this presupposes that atoms are like billiard balls with distinguishing marks and permanent identities. They aren't. Two atoms of the same kind are indistinguishable. One cannot 'put labels on them' and recognize them individually later. Moreover, at the deeper, subatomic level the atoms themselves are in a perpetual state of flux. We think things persist in time because structures persist, and we mistake the structure for substance. But looking for enduring substance is like looking for time. It slips through your fingers. One cannot step into the same river twice.

Zeno of Elea, who belonged to the same philosophical school as Parmenides, formulated a famous paradox designed to show that motion is impossible. After an arrow shot at a target has got halfway there, it still has half the distance to go. When it has gone half that distance, it still has half of that way to go. This goes on for ever. The arrow can never reach the target, so motion is impossible. In normal physics, with a notion of time, Zeno's paradox is readily resolved. However, in my timeless view the paradox is resurrected, but the arrow never reaches the target for a more basic reason: the arrow in the bow is not the arrow in the target.

There are two parts to my claim that time does not exist. I start from the philosophical conviction that the only true things are complete possible configurations of the universe, unchanging Nows. Unchanging things do not travel in time from Now to Now. Material things, we included, are simply parts of Nows. This philosophical standpoint must be matched by a physical theory that seems natural within it. The evidence that such a physical theory exists and seems to describe the universe forms the other part of my claim. This section has merely made the philosophy, the notion of being, clear. The physics, the guts of the story, is still to come.

THE BIG PICTURE

Before Newton was born, René Descartes raised a nightmarish prospect. How do I know, he asked, whether anything exists? Is some malignant demon conjuring up my thoughts and experiences? Perhaps there isn't any world. How can we be sure of anything? Descartes famously argued that we can at least be certain of our own existence. *Cogito ergo sum*: I think, therefore I am. In fact, this did not get him very far, and his main argument for a real world was that God would not deceive us on such a fundamental matter.

Modern science has a better answer to the solipsists – those who, like Descartes in that extreme moment of doubt, deny existence outside their own thoughts. The starting point is that we do observe a great variety of phenomena. We can then ask whether we can postulate a world and laws that lead to the phenomena. If this is so, it does not explain how or why the world is there, but it does provide grounds for taking its existence more seriously.

You may think that time capsules and a brain preserved in aspic aware of seeing motion are getting dangerously close to solipsism and the machinations of a demon. Without anticipating the rest of the book, an outline may still be helpful. There are only two rules of the game: there must be an external world subject to laws and a correspondence between it and experiences.

Apart from the fact that Newton placed the material objects of the universe in an arena, my things are his things. They are Nows, the relative configurations of the universe. Newton's Nows form a string, brought into being by an act of creation at one end, called the past. It is usually assumed that our experiences in some instant reflect the structure in a short segment of the string at a point along it. It is a segment, rather than one Now, because we see things not only in positions but in motion. However, a single Now contains only positional information. It seems that we need at least two Nows to have information about changes of position.

Newtonian history, as modified by Big Bang cosmology, translates into a path in Platonia. It begins at a certain point with a creation event, after which the laws of nature determine the path. Many paths satisfy the same laws, but the laws by themselves do not tell us why one path is chosen by the creation event in preference to others.

The alternative picture, suggested by quantum mechanics and proposed in this book, is quite different. There are no paths with unique starting points conceived as creation events. Indeed, there are no paths at all. Instead, the different points of Platonia, each of which represents a different possible configuration of the universe, are present – as potentialities at least – in different quantities. This matches what we found in the Timeless Theory bag: many different triangles present in different quantities. It will be helpful to represent this in a more graphic way. Imagine that Platonia is covered by a mist. Its intensity does not vary in time – it is static – but it does vary from position to position. Its intensity at each given point is a measure of how many configurations (as in the previous example, with triangles in the Timeless Bag) corresponding to that point are present. All these configurations, present in different quantities, you should imagine for the moment as being collected together in a 'heap' or 'bag'.

So, Platonia is covered with mist. Its intensity cannot change in time (there is no time), but it does vary from point to point. In some places it is much more intense than in others. A timeless law, complete in itself, determines where the mist collects. The law is a kind of competition for the mist between the Nows. Those that 'resonate' well with each other get more mist. The outcome is a distribution of mist intensity. This, as I have just explained, is simply another description of the Timeless Theory bag – for mist intensity read numbers of triangle copies. But the Nows of this Platonia are much more complex than triangles.

This opens up possibilities. Triangles tell no stories, they are too simple. But if the Nows are defined by, say, the arrangements of three large bodies and of many thousands of small bodies, things are different. For example, the three large bodies could form the tenth triangle from the right in Figure 1. The remaining small bodies could be arranged in such a way that they literally create the pattern of the first nine triangles from the right of the sequence. This may seem contrived, but it is possible. It is a Now in a greatly enlarged Platonia. Shown such a Now, what could we make of it? One interpretation is that the small bodies record what the large bodies have done: the Now is a time capsule, a picture of a Newtonian history. As soon as a sufficient number of bodies are present, the possibilities for creating time capsules are immense.

I believe the sole reason we believe in time is because we only ever experience the universe through the medium of a time capsule. My assumptions are:

(1) All experience we have in some instant derives from the structure in one Now.

(2) For Nows capable of self-awareness (by containing brains, etc.) the *probability* of being experienced is proportional to their mist intensity.

(3) The Nows at which the mist has a high intensity are time capsules (they will also possess other specific properties).

Thus, the one law of the universe that determines the mist intensity over Platonia is timeless. The Nows and the distribution of the mist are both static. The appearance of time arises solely because the mist is concentrated on time capsules, and a Now that is a time capsule is therefore much more likely to be experienced than a Now that is not. (Please remember that this is only an outline: the detailed arguments are still to come.)

Of the three assumptions, the second is the most problematic. The first and third may seem strange and implausible, but they can be made definite. If correct, their significance and meaning are clear-cut. Both could be shown to be false, but this is good, since a theory that cannot be disproved is a bad theory. The best theories make firm predictions that can be tested. The main difficulty with the second assumption is in saying what it means. We encounter, in a modified form, the difficulty that Descartes raised. It is acute.

In a Newtonian scheme, the connection between theory and experience is unambiguous. There is a path through Platonia, and all the Nows on it are realized: sentient beings within any Nows on the path do experience those Nows. In the alternative scheme, the distribution of the mist over Platonia – its intensity at each Now – is as definite as the line of the Newtonian path. The difficulty, which is deeply rooted in quantum mechanics, is how to interpret the intensity of the mist. When we get to grips with quantum mechanics, I shall explain my reasons for assuming that the mist intensity at a Now measures its probability of being experienced. Perhaps some cosmic lottery is the best way to explain this.

Each Now has a mist intensity. Suppose that all the Nows participate in a lottery, receiving numbers of tickets proportional to their mist intensities. Nows where the mist is intense get tickets galore, others very few. By assumption (1), conscious experience is always in one Now. If a Now has a special structure, it is capable of self-awareness. But is it actually self-aware? Structure in itself, no matter how intricate and ordered, cannot explain how it can be self-aware. Consciousness is the ultimate mystery.

Perhaps it is a mystery that makes some sense of the mist that covers

Platonia. If there is a cosmic lottery, clearly the Nows with the most tickets will have the best chance. If a ticket belonging to a Now capable of self-awareness is drawn, this can, so to speak, 'bring to life' the Now. It is aware. The consciousness potentially present in Nows structured the right way is actual in those that are drawn. Two questions about this cosmic lottery may well be asked: when are the tickets drawn, and how many are drawn?

The first question is easily answered: it has no meaning. Think of the brain preserved in aspic, or the unfortunate brain-damaged patient who believes that Harold Macmillan is Prime Minister and Dwight Eisenhower is President. The structure capable of making a Now self-aware is eternal and timeless. Structure is all that counts. Self-awareness does not happen at a certain time and last for some fraction of a second. Yesterday seems to come before today because today contains records (memories) of yesterday. Nothing in the known facts is changed by imagining them hung on a 'line of time' – or even reversing their positions on that line. The instant is not in time, time is in the instant. We do not have to worry when the draw is made, only whether our number comes up.

The question of how many tickets are drawn is a tough one. If only one is drawn, your present Now, which does exist, must be the one and only instant realized and experienced. All your memories are then illusions in the sense that you never experienced them. That seems very hard to believe. What is more, memories are legion. If you believe you did actually experience them all, then lots of Nows have been drawn. From this it is a small step to saying that all Nows in Platonia are drawn. In quantum mechanics, this is called the many-worlds hypothesis. But then the theory seems to become vacuous: everything that can be is, no predictions appear to be made. The root of the problem is the assumption, neat and clean in itself, that each experienced instant is always tied to a single Now and that the distribution of the mist over Platonia is determined by a law indifferent to the workings of the cosmic lottery. Whether or not particular Nows are drawn has no effect on the mist intensity. The rules of the scheme make it quite impossible to say how many, if any, of your memories are real. All we know is that the present Now is real. You can see how Descartes's dilemma is revived in such a scheme. I suspect that it is a problem we just have to live with.

The theory is still testable because only Nows with high mist intensity (and therefore high probability) are likely to be experienced, and such

Nows have characteristic properties: above all, they are time capsules. We can therefore test our own experiences and see if they verify the predictions of the theory. This is something that in principle can be settled by mathematics and observations. For if physicists can determine or guess the structure of Platonia and formulate the law that determines how the mist is distributed over it, then it is simply a matter of calculation to find out where in Platonia the mist is most intense. If the mist is indeed concentrated on structures that are time capsules, the theory will make a very strong prediction – any Now that is experienced will contain structures that seem to be records of a past of that Now. It will also contain other characteristic structures.

The huge number of things that can coexist simultaneously in one Now is significant here. It means that many independent tests can be made on a single time capsule to see whether the predictions are confirmed. The laws of nature are usually tested by repeating experiments in time. If the same initial state gives the same outcome, the law is confirmed. However, for an object as richly structured as the Earth (which in any instant belongs to one of the Nows in Platonia), repeating experiments in time can be replaced by repeating them in space. As it happens, even confirming a theory by repeating experiments in time as normally understood boils down to comparing records in one Now. The precondition of all science is the existence of time capsules. All the Nows we experience are time capsules. The question is whether we can explain why this is so from first principles: can the strong impression of time emerge from timelessness? It is a logical possibility, but the real test must await mathematical advances. Unfortunately, they are not likely to be easy.

Strange as a timeless theory may seem, it has the potential to be very powerful. Boltzmann's work highlighted two difficulties inherent in any theory of time – initial conditions must be imposed arbitrarily; and dull, unstructured situations are far more probable than the interesting structured things we find all around us. Interestingly structured Nows are an extreme rarity among all the Nows that can be. If the mist does pick out time capsules in Platonia, it must be very selective. Since all possible structures are present in Platonia, the vast majority of Nows do not contain any structures at all that could be called records. Even then, the apparent records will be mutually consistent in only a tiny fraction of what is already a tiny fraction. Only our habitual exposure to the time capsules we experience blinds us to the magnitude of the phenomenon that needs to be explained. Stars in real space give us only an inkling of how thinly time capsules are spread. Any scheme that does select them will be very powerful. But more than that, it will be more fully rational than classical physics, with its need to invoke a very special initial condition, can ever be. Once the law that governs the distribution of the mist over Platonia has been specified, nothing more remains to be done. The mist gathers where it does for only two reasons: the structure of the law and the structure of Platonia.

So where is the mist likely to gather? The mathematics needed to answer this question will certainly be difficult, but there are some hints (which I shall elaborate in the final chapters). They suggest that mist is likely to be distributed along thin, gossamer-like filaments that bifurcate and form a tree-like structure (Figure 6).

A tendency to bifurcation is deeply rooted in quantum mechanics. In principle, it could happen in both directions along a filament. However, the Nows we experience all seem to have arisen from a unique past. There seems to be no branching in that direction. Within quantum mechanics, as presently formulated in space and time, this fact is not impossible, but it is as puzzling as the low entropy that so exercised Boltzmann. It does seem improbable. I suspect that everything will look different if we learn to think about quantum mechanics in Platonia. For one thing, the arena has a very different shape. This is why I was keen to show you at this early stage the diagrams of Triangle Land (Figures 3 and 4) and my representation of Platonia (Figure 5). It opens out in one direction from nothing. I suspect that the branching filaments of mist in Figure 6 arise because they reflect this overall, flower-like structure of Platonia. If that is so, the great asymmetries of our existence – past and future, birth and death – arise from a deep asymmetry in being itself. The land of possible things has one absolute end, where it abuts onto mere nothing, but it is unbounded the other way, for there is no limit to the richness of being.

Who knows what experiences are possible in the oases of richly structured Nows strung out along the trade routes that cross the deserts of Platonia? The plurality of experience is remarkable and suggestive. In any instant, we are aware of many things at once. Through memories we are, as it were, present simultaneously in many different Nows in Platonia. Richness of structure permits this. One grand structure contains substructures that are 'pictures' – simplified representations that capture the essential features – of other structures. Our memories are pictures of other

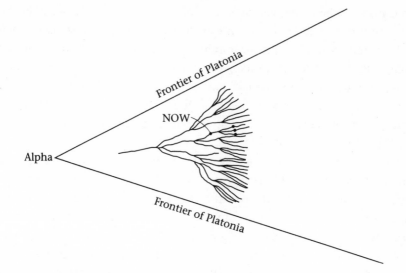

Figure 6 The conjectured filamentary distribution of mist in Platonia. The instant you experience now is marked NOW. To its left lie Nows of which you have memories in NOW. There is no bifurcation in this direction, matching our conviction that we have a unique past. In the other direction there is a branching into different alternative 'futures' of NOW. In all of them, you think you have advanced into the future by the same amount from NOW. These different filaments are 'parallel worlds' that seem to have a common past, to which NOW belongs. Note that the filaments have a finite width, unlike a Newtonian string of successive instants. All around NOW, along the filament and to either side of it, are other Nows with slightly different versions of yourself. All such Nows are 'other worlds' in which there exist somewhat different but still recognizable versions of yourself. In other filaments are worlds you would not recognize at all.

Nows within this Now, rather like snapshots in an album. Each Now is separate and a world unto itself, but the richly structured Nows 'know' about one another because they literally contain one another in certain essential respects. As consciousness surveys many things at once in one Now, it is simultaneously present, at least in part, in other Nows. This awareness of many things in one could well exist in a much more pronounced form in other places in Platonia.

The picture of ourselves dividing into parallel Nows may be unsettling, but the phenomenon itself is familiar. We are used to being in different Nows and being slightly different in all of them – that is simply the effect of time as it is usually conceived. The account of Lucy's leaps emphasized that the differences in ourselves between Nows are far greater than we realize within consciousness. Huge numbers of microscopically different

Nows could give identical conscious experience. As we shall see, quantum mechanics forces us to consider Nows everywhere, not just those on one path. It unsettles by division, seeming to threaten dissolution and personal integrity. But it simultaneously binds us into the far mightier whole of everything that can be, doing so much more decisively than any Newtonian scheme can do. For the Nows that are likely to be experienced are the ones that are most sensitive to the whole of Platonia.

I think this is sufficient introduction. I could go on to talk about free will, the future, our place in the universe, religion, and so on. If the theory is correct, it must change the way we think about these things. However, without some real understanding of the arguments for a timeless universe, I feel further discussion would lack a solid basis. I therefore postpone these issues to later in the book, especially the epilogue. My aim so far has been to outline the scheme and to show that it is truly timeless and at least logically possible.

The Invisible Framework and the Ultimate Arena

Newton introduced two 'great invisibles' as the arena of physics: absolute space and time. In Part 2 we shall see why they have appeared for so long to be better suited to acting as the frame of the world than Platonia. It is all to do with an issue that physicists and philosophers have been arguing about for centuries: is motion absolute or relative? Newton's position has seemed to be so strong that many people still believe it cannot be overthrown. But it can. The demonstration of the relatively simple solution in Newtonian physics will prepare us for the almost miraculous way in which things work out in Einstein's theory (Part 3). They give the strongest suggestion that quantum cosmology – and hence our universe – is timeless. That we come to in Parts 4 and 5. Chapter 4 is a brief historical introduction, and sets the scene for the remainder of Part 2 – and for much of the rest of the book.

CHAPTER 4

Alternative Frameworks

ABSOLUTE OR RELATIVE MOTION?

Both Copernicus and Kepler believed that the universe, with the solar system at its centre, was bounded by a huge and distant rigid shell on which the luminous stars were fixed. They did not speculate what lay beyond – perhaps it was simply nothing. They defined all motions relative to the shell, which thus constituted an unambiguous framework. Many factors, above all Galileo's telescopic observations in 1609 and the revival of interest in the Greeks' idea of atoms that move in the void, destroyed the old cosmology. New ideas crystallized in a book that Descartes wrote in 1632. He was the first person to put forward clearly an idea which, half a century later, Newton would make into the most basic law of nature: if nothing exerts a force on them, all bodies travel through space for ever in a straight line at a uniform speed. This is the law of inertia. Descartes never published his book because in 1633 the Inquisition condemned Galileo for teaching that the Earth moves. The Copernican system was central to Descartes's ideas, and to avoid Galileo's fate he suppressed his book.

He did publish his ideas in 1644, in his influential *Principles of Philosophy*, but with a very curious theory of relative motion as an insurance policy. He argued that a body can have motion only relative to some other body, chosen as a reference. Since any other body could play the role of reference, any one body could be regarded as having many different motions. However, he did allow a body to have one true 'philosophical motion', which was its motion relative to the matter immediately adjacent to it. (Descartes believed there was matter everywhere, so any body did always have matter adjacent to it.) This idea let him off the Inquisition's hook, since he claimed that the Earth was carried around the Sun in a huge vortex, as in a whirlpool. Since the Earth did not move relative to the immediately adjacent matter of the vortex, he argued that it did not move!

However, he then formulated the law of inertia, just as in 1632. When, sometime around 1670, long after Descartes's death in 1650, Newton came to study his work, he immediately saw the flaw. To say that a body moves in a straight line presupposes a fixed frame of reference, which Descartes had denied. Since Newton could see the great potential of the law of inertia, to exploit it he came up with the concept of an immovable space in which all motion takes place. He was very scornful of Descartes's inconsistency, and when he published his own laws in 1687 he decided to make it a big issue, without, however, mentioning Descartes by name. He introduced the notion of absolute space and, with it, absolute time.

Newton granted that space and time are invisible and that one could directly observe only relative motions, not the absolute motions in invisible space. He claimed that the absolute motions could nevertheless be deduced from the relative motions. He never gave a full demonstration of this, only an argument designed to show that motion could not be relative. He was making a very serious point, but at the same time he wanted to make a fool of Descartes. This had strange and remarkable consequences.

Descartes had sought to show that all the phenomena of nature could be explained mechanically by the motion of innumerable, tiny, invisible particles. Vital to his scheme was the centrifugal force felt as tension in a string that retains a swung object. The object seems to be trying to escape, to flee from the centre of rotation. In Newtonian terms, it is actually trying to shoot off along the tangent to the circle, but that is still a motion that would take it away from the centre and create the tension. Descartes claimed that light was pressure transmitted from the Sun to the Earth by centrifugal tension set up in the vortex that he pictured swirling around the Sun. Because centrifugal force was so important to Descartes, Newton used it to show that motion cannot be relative. Newton's intention was to hoist Descartes by his own petard.

Newton imagined a bucket filled with water and suspended by a rope from the ceiling. The bucket is turned round many times, twisting the rope, and is then held still until the water settles. When the bucket is released, the rope unwinds, twisting the bucket. Initially the surface of the water remains flat, but slowly the motion of the bucket is transmitted to the water, which starts to spin, feels a centrifugal force and starts to rise up the side of the bucket. After a while, the water and bucket spin together without relative motion, and the water surface reaches its greatest curvature.

Newton asked what it was that caused the water's surface to curve. Was it the water's motion relative to the side of the bucket (Descartes's claimed true philosophical motion relative to the immediately adjacent matter) or motion relative to absolute space? Surely the latter, since when the relative motion is greatest, at the start, there is no curvature of the water's surface, but when the relative motion has stopped (and the water and bucket spin together) the curvature is greatest. This was Newton's main argument for absolute space. It was strong and it ridiculed Descartes.

In Newton's lifetime, his notion of absolute space, to which he gave such prominence, attracted strong criticism. If space were invisible, how could you say an object moves in a straight line through a space you cannot see? Newton never satisfactorily answered this question. Many people felt, as Descartes did, that motion must be relative to other matter, though not necessarily adjacent matter. Bishop Berkeley argued that, as in Copernican astronomy, motion must ultimately be relative to the distant stars, but he failed to get to grips with the problem that the stars too must be assumed to move in many different ways and thus could not define a single fixed framework, as Copernicus and Kepler had believed.

Newton's most famous critic was the great German mathematician and philosopher Wilhelm Gottfried Leibniz, who had been involved in a very unpleasant dispute with Newton about which of them had first discovered the calculus, the revolutionary new form of mathematics that made so many things in science much easier, including the development of mechanics. In 1715, Leibniz began a famous correspondence on Newton's ideas with Samuel Clarke, who was advised by Newton. The *Leibniz–Clarke Correspondence* has become a classic philosophy text. Many undergraduates study it, and philosophers of science often discuss it.

The exchange had an inconclusive outcome. It is generally agreed that Leibniz advanced effective philosophical arguments, but he never addressed the detailed issues in mechanics. Typically, he argued like this. Suppose that absolute space does exist and is like Newton claimed, with every point of space identical to every other. Now consider the dilemma God would have faced when he created the world. Since all places in absolute space are identical, God would face an impossible choice. Where would he put the matter? God, being supremely good and rational, must always have a genuine reason for doing something – Leibniz called this the 'principle of sufficient reason' (I have already appealed to this when I discussed brain function and consciousness, by requiring an observable effect to have an observable cause) – and because absolute space offered no distinguished locations, God would never be able to decide where to put the matter. Absolute time, on the assumption that it existed, presented the same difficulty. Newton had said that all its instants were identical. But then what reason could God have for deciding to create the world at some instant rather than another? Again, he would lack a sufficient reason. For reasons like these, not all of them so theological, Leibniz argued that absolute space and time could not exist.

A century and a half passed before the issue became a hot topic again. This raises an important issue: how could mechanics have dubious foundations and yet flourish? That it flourished nevertheless was due to fortunate circumstances that are very relevant to the theme of this book. First, although the stars do move, they are so far away that they provide an effectively rigid framework for defining motions as observed from the Earth. It was found that in this framework Newton's laws do hold. It is hard to overestimate the importance of this fortunate effective fixity of the distant stars. It presented Newton with a wonderful backdrop and convenient framework. Had the astronomers been able to observe only the Sun, Moon and planets but not the stars (had they been obscured by interstellar dust), Newton could never have established his laws. Thus, scientists were able to accept Newton's absolute space as the true foundation of mechanics, using the stars as a substitute for the real thing – that is, a true absolute frame of reference. They also found that Newton's uniformly flowing time must march in step with the Earth's rotation, since when that was used to measure time (in astronomical observations spanning centuries, and even millennia) Newton's laws were found to hold. Once again, a substitute for the 'real thing' was at hand. One did not have to worry about the foundations. Fortunate circumstances like these are undoubtedly the reason why it is only recently that physicists have been forced to address the issue of the true nature of time.

The person who above all brought the issue of foundations back to the fore was the Austrian physicist Ernst Mach, whose brilliant studies in the nineteenth century of supersonic projectiles and their sonic boom are the reason why the Mach numbers are named after him. Mach was interested in many subjects, especially the nature and methods of science. His philosophical standpoint had points in common with Bishop Berkeley, but even more with the ideas of the great eighteenth-century Scottish empiricist David Hume. Mach insisted that science must deal with genuinely observable things, and this made him deeply suspicious of the concepts of invisible absolute space and time. In 1883 he published a famous history of mechanics containing a trenchant and celebrated critique of these concepts. One suggestion he made was particularly influential.

It arose as a curious consequence of the covert way Newton had attacked Descartes. Considering Newton's bucket argument, Mach concluded that, if motion is relative, it was ridiculous to suppose that the thin wall of the bucket was of any relevance. Mach had no idea that Newton was attacking Descartes's notion of the one true philosophical motion, just as Newton had not seen that Descartes had invented it only to avoid the wrath of the Inquisition. Newton had used the bucket argument to show that relative motion could not generate centrifugal force, but Mach argued that the relative motions that count are the ones relative to the bulk of the matter in the universe, not the puny bucket. And where is the bulk of the matter in the universe? In the stars.

This led Mach to the revolutionary suggestion that it is not space but all the matter in the universe, exerting a genuine physical effect, that creates centrifugal force. Since this is just a manifestation of inertial motion, which Newton claimed took place in absolute space, Mach's proposal boiled down to the idea that the law of inertia is indeed, as Bishop Berkeley believed, a motion relative to the stars, not space. Mach's important novelty was that there must be proper physical laws that govern the way distant matter controls the motions around us. Each body in the universe must be exerting an effect that depends on its mass and distance. The law of inertia will turn out to be a motion relative to some average of all the masses in the universe. For this basic idea, Einstein coined the expression *Mach's principle*, by which it is now universally known (though attempts at precise definition vary quite widely).

Mach's idea suggests that the Newtonian way of thinking about the workings of the universe, which is still deep-rooted, is fundamentally wrong. The Newtonian scheme describes an 'atomized' universe. The most fundamental thing is the containing framework of space and time: that exists before anything else. Matter exists as atoms, tiny unchanging masses that move in space and time, which govern their motion. Except when close enough to interact, the atoms move with complete indifference to one another, each following a straight and lonely path through the infinite reaches of absolute space. The Machian idea takes the power from space and time and gives it to the actual contents of the universe, which all dance in their motions relative to one another. It is an organic, holistic view that knits the universe together. Very characteristic is this remark of Mach in his *The Science of Mechanics* (pp.287–8):

> Nature does not begin with elements, as we are obliged to begin with them. It is certainly fortunate for us that we can, from time to time, turn aside our eyes from the overpowering unity of the All and allow them to rest on individual details. But we should not omit, ultimately to complete and correct our views by a thorough consideration of the things which for the time being we have left out of consideration.

Mach himself made only tentative suggestions for a new relative mechanics, but his remarks caught the imagination of many people, above all Einstein, who said that Hume and Mach were the philosophers who had influenced him most deeply. Einstein spent many years trying to create a theory that would embody Mach's principle, and initially believed that he had succeeded in his general theory of relativity. That is why he gave it that name. However, after a few years he came to have doubts. Eventually he concluded that Mach's idea had been made obsolete by developments in physics, especially the theory of electromagnetism developed by Faraday and Maxwell, which had introduced new concepts not present in Newton's scheme.

Throughout the twentieth century, physicists and philosophers discussed Mach's principle at great length, without coming to any conclusion. It is my belief that the problem lies in Einstein's highly original but indirect approach. Mach had not made a really clear proposal, and Einstein never really stopped and asked himself just what should be achieved by Mach's principle. I shall consider this in Part 3, but I need to anticipate a small part of the story in order to justify Part 2. Einstein's theory is rather complicated and achieves several things at once. It is not easy to separate the parts and see the 'Machian' structure. In my opinion, general relativity is actually as Machian as it could be. What is more, it is the Machian structure that has such dramatic consequences when one tries to reconcile the theory with quantum mechanics. If, as I believe, the quantum universe is timeless, it is so because of the Machian structure of general relativity. To explain the core issues, I need a simplified model that captures the essentials. This Part 2 will provide. It will also provide a direct link between the great early debate about the foundations of mechanics and the present crisis of quantum cosmology. Two key issues are still the same: what is motion, and what is time? It will also enable me to explain the main work in physics with which I have been involved, and make it easier for you to see why I have come to doubt the existence of time.

Science advances in curious ways, and scientists are often curiously unconcerned with foundations. Descartes was one of the greatest philosophers, yet in that first book in 1632 he never gave a moment's thought to the definition of motion. We are so used to living on the solid Earth that it seems unproblematic to say that a body moves in a straight line. If the Inquisition had not condemned Galileo, Descartes would never have argued for the relativity of motion. But for the inconsistency of his system, Newton would not have made an issue out of absolute space and time. He would not have devised the bucket argument, Mach might never have had his novel idea, and Einstein would not have been inspired to his greatest creation.

Had the Inquisition condemned Galileo a few months later, Descartes would have published his ideas in their original form – and general relativity might never have been found.

AN ALTERNATIVE ARENA

I would like to say a bit more about my own personal development, which as the book progresses will help you to understand why I am so deeply convinced of the need to have a new concept of time. In the very first days after my trip to the Bavarian Alps, while thinking hard about time, I came across Mach's book. Like so many others, I was captivated by his idea about inertia. His comments on time also encouraged me greatly: 'It is utterly beyond our power', he said, 'to measure the changes of things by time. Quite the contrary, time is an abstraction, at which we arrive by means of the changes of things.' This was just the conclusion I had reached. A year or so later, after I had decided to study the foundations of physics, I started to read the papers Einstein had written when he was creating general relativity. Comparing them with what Mach had written, I came to the conclusion that Einstein had simply not set about the problem in the right way: he had not attacked it directly. It seemed to me necessary to go back to first principles.

It was six or seven years before I came to form really clear ideas. I eventually concluded that what was needed above all was a new arena in which to describe the universe. I arrived at the notion of Platonia (or, as I originally called it, the relative configuration space of the universe). The argument was quite simple. First, it is a fact that we orient ourselves in real life by objects we actually see, not by invisible space (see the Notes on the previous chapter). Things are the signposts that tell us where we are. There is also the fortunate fact that we live on the nearly rigid Earth. We can orient ourselves by means of just a few objects fixed on its surface, say church spires when hiking in the English countryside. Always there, the Earth provides a natural background. Motion seems to take place in a framework. But imagine what life would be like if we lived on a jellyfish!

The fact is that we live in a very special location. Only the tiniest fraction of matter in the solar system, let alone the universe, is in solid form. Imagine that we lived in an environment much more typical of the universe – in space. To simplify things, let there be only a finite number of objects, all in motion relative to one another. At any instant there are certain distances between these other objects and us. There is nothing else. In these circumstances, what would be the natural way to answer what is always a fundamental question: where are we? We have no other means of saying where we are except in terms of our distances to other objects. What is more, it would be artificial to choose just a few of them to locate ourselves. Why these rather than those? It would be much more natural to specify our distances to *all* objects. They define our position. This conclusion is very natural once we become aware that nothing is fixed. Everything moves relative to everything else.

Taking this further, thinking about the position and motion of one object is artificial. We are part of Mach's All, and any motion we call our own is just part of a change in the complete universe. What is the reality of the universe? It is that in any instant the objects in it have some relative arrangement. If just three objects exist, they form a triangle. In one instant the universe forms one triangle, in a different instant another.

What is to be gained by supposing that either triangle is placed in invisible space? The proper way to think about motion is that the universe as a whole moves from one 'place' to another 'place', where 'place' means a relative arrangement, or configuration, of the complete universe.

An arena is the totality of places where one can go in some game. But who is playing the game and where? In Newton's game, individual objects play in absolute space. In Mach's game, there is only one player – the universe. It does not move in absolute space, it moves from one configuration to another. The totality of these places is its relative configuration space: Platonia. As the universe moves, it therefore traces out a path in Platonia. This captures, without any redundant structure, the idea of history. History is the passage of the universe through a unique sequence of states. In its history, the universe traces a path through Platonia.

However, such language makes it sound as though time exists. I may have inadvertently conjured up an image in your mind of the universe as a lone hiker walking the fells in northern Platonia. Properly understood, the Machian programme is much more radical. For no Sun rises or sets over that landscape to mark the walker's progress. The Sun, like the moving parts of any clock, is part of the universe. It is part of the walker. Of course, to say that time has passed, we must have some evidence for that. Something must move. That is the most primitive fact of all. In the Newtonian picture, as in Feynman's quip, time can pass without anything happening. If we deny that, the grandstand clock must go. There is nothing outside the universe to time it as it goes from one place to another in Platonia – only some internal change can do that. But just as all markers are on an equal footing for defining position, so are all changes for the purposes of timing. We must reckon time by the totality of changes. But changes are just what takes the universe from one place in Platonia to another. Any and all changes do that. We must not think of the history of the universe in terms of some walker on a path who can move along it at different speeds. The history of the universe *is* the path. Each point on the path is a configuration of the universe. For a three-body universe, each configuration is a triangle. The path is just the triangles – nothing more, nothing less.

With time gone, motion is gone. If you saw a jumbled heap of triangles, it would not enter your head that anything moved, or that one triangle changed into another. When Newton's superstructure is removed,

Newtonian history is like that jumbled heap of triangles, except that it is a special heap. If you picked up each triangle – I call that picking up an instant of time – and marked its position in Triangle Land, you would find that the marks of the triangles form a continuous curve.

This was the decisive picture that crystallized in my mind about 1971. At that stage I had no thought of applications to quantum mechanics, and no inkling that it might lead to the replacement of one clearly delineated path through Platonia by a mist that hovers over the same timeless landscape. We had a blackboard in our kitchen in College Farm, and I wrote at the top it it: 'The history of the universe is a continuous curve in its relative configuration space.' My wife, perhaps understandably, was rather sceptical about the progress I was making. After all, fourteen words were not much to show for seven years of thought. But the clear formulation of the concept of Platonia was the important thing. It shifts attention from the parts of the universe to the universe itself. It shows that time is not needed as an extra element, the Great Timekeeper outside the universe. The universe keeps track of itself. In one instant it is where it is, in another it is somewhere else. That is what a different instant of time is: it is just a different place in Platonia. Instants of time and positions of the objects within the universe are all subsumed into the single notion of place in Platonia. If the place is different, the time is different. If the place is the same, time has not changed. This change of viewpoint is made possible only because the universe is treated as a single whole and time is reduced to change.

I think the reason why I take the possibility of a completely timeless universe more seriously than almost all other physicists is this background that came from thinking about Mach's principle. As we shall see, Platonia is the natural arena for the realization of that idea. Many years after I had first recognized that Platonia would provide the basis for the solution to the Machian problem, I began to see that it had deep relevance in the quantum domain too. The problems of the origin of inertia and of quantum cosmology form a seamless whole.

CHAPTER 5

Newton's Evidence

THE AIMS OF MACHIAN MECHANICS

Merely changing the framework in which one conceives of the universe does nothing, but it is still very illuminating to look at some fundamental facts of mechanics in the alternative arenas of absolute space and Platonia. This exercise brings out the strengths of Newton's position, and at the same time shows what a Machian approach must achieve. The following discussion is based on penetrating remarks made in 1902 by the great French mathematician Henri Poincaré. More clearly than Mach, he demonstrated what is required of a theory of relative motion. Unfortunately, his remarks were overshadowed by Einstein's discovery of relativity and did not attract the attention they deserved – and still deserve.

You may find that this chapter requires more reflection than all the others. You certainly do not need to grasp it all, but I hope that you will be able to change from a way of thinking to which we have been conditioned by the fact that we evolved on the stable surface of the Earth to a more abstract way of thinking that would have been forced upon us had we evolved from creatures that roamed in space between objects moving through it in all directions. We have to learn how to find our bearings when the solid reassuring framework of the Earth is not there. This is the kind of mental preparation you need to understand the ideas Poincaré developed. In this respect, he was smarter than Einstein.

Poincaré simply asked, rather more precisely than anyone before him, what information is needed to predict the future. Another French mathematician, Pierre Laplace, had already imagined a divine intelligence that

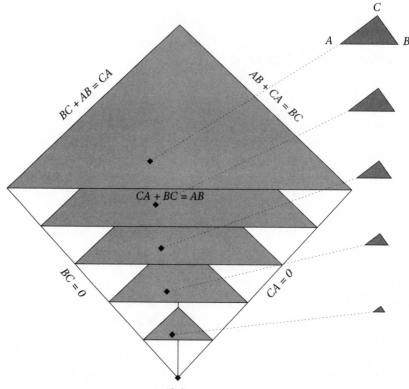

Alpha

Figure 7 Here Alpha, the apex of Triangle Land, is at the bottom. The axes that were shown in Figures 3 and 4 have been removed since they would detract from the essence of this and the following figure. Two of the ribs of Triangle Land run upwards to the right and left. They are marked $BC = 0$ and $CA = 0$ to indicate that the triangles corresponding to points on these ribs have 'collapsed' because their sides BC and CA, respectively, are zero. The third rib recedes into the figure (this is the rib that in Figure 4 runs along the 'floor'). The shaded planes cut the faces of the pyramid in the lines on the 'sheets' in Figure 4. As shown here, all points on any straight line from Alpha through Triangle Land represent different triangles, but they differ only in size. In fact, all the points on any one of the shaded sheets represent triangles that have the same perimeter. If we are interested only in the *shapes* of the triangles, these are represented by the points on just one of the planes (i.e. the different points on any one plane represent differently shaped triangles; this possibility is depicted more fully in Figure 8).

at one instant knows the positions and motions of all bodies in the universe. Using Newton's laws, the divinity can then calculate all past and future motions – it can see, in its mind's eye, all of history laid out for the minutest inspection. As an alternative to the standard representation in Newton's absolute space, it will help to see this miracle performed in

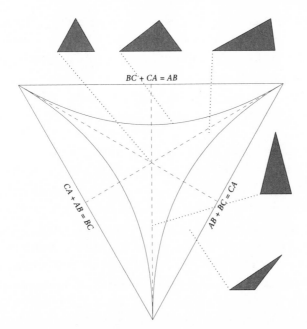

$$BC + CA = AB$$

$$CA + AB = BC$$

$$AB + BC = CA$$

Figure 8 This shows one of the shaded planes in Figure 7. It can be called Shape Space, because each point in it represents a different possible shape of a triangle. The point at the centre represents an equilateral triangle (all three sides equal). Points on the three dashed lines correspond to isosceles triangle (two sides equal). All other possible triangles are scalene (all three sides unequal). Points on the three curved lines correspond to right-angled triangles (the central triangle at the top is the Pythagorean 3, 4, 5 triangle). All points inside the curved lines correspond to acute triangles (all angles less than 90°); all points outside the curved lines correspond to obtuse triangles (one angle greater than 90°). When I asked my friend Dierck Liebscher to create this diagram, I had no idea that it would turn out to be so beautiful. I do like the curved lines of the right-angled triangles! Let me remind you that all the points on the straight edges of Shape Space correspond to triangles that have become 'flat' because all three corners of each of these triangles are collinear (on one line). Each of the three vertices of Shape Space correspond to configurations in which two particles coincide, while the third is some distance from them.

Triangle Land, the simplest Platonia. This will reveal a curious defect in Newtonian mechanics.

You may like to refresh your memory by returning to Figures 3 and 4 before you examine Figures 7 and 8. Figures 3, 4, 7, 8 and 9 are very important. I am rather concerned that younger readers (those under forty, or even fifty!) may have some difficulty with them, since fundamental geometry is not taught nearly as thoroughly at school as it used to be. However, if you can spend a bit of time on these figures and begin to understand what they mean, you will certainly get a great deal more out of this book. In fact, you will also be absorbing some of the deepest and most fruitful concepts in mathematics and theoretical physics. Don't worry – it can be done. Once the clutter of technical detail is removed, all the great ideas in mathematics and physics are in essence very simple and intuitive. But you need patience to absorb them. When Newton was asked how he had come to make his great discoveries, he answered: 'By thinking about these things for a long time.' Try lying in bed or a nice warm bath and thinking about Triangle Land!

Figure 8 brings out the rich topography of any Platonia (Shape Space is a possible Platonia). Wherever you go, you find something different. Each point is a different 'world' – and a different instant of time. There are even characteristically different regions (of acute and obtuse triangles), like provinces or counties, as well as internal and external frontiers (the right-angled and isosceles triangles). Any Platonia is quite unlike Newton's absolute space, all points of which are identical. As I remark in the Notes to Chapter 4, there is something unreal about that property of absolute space. Real things have genuine attributes that distinguish them from other real things. Platonia is a land of real things. I find Figure 8 very suggestive. Leibniz always said that it is necessary to consider all possible worlds and find some reason why one rather than another occurs or is actually created. In Figure 8, we do see all the possible worlds of triangle shapes laid out before us. Box 3 contains a short digression on possible kinds of Platonias.

BOX 3 Possible Platonias

One of the big unsolved problems of physics is the origin of distance and whether it is absolute. Since we need a measuring rod to measure any distance, this suggests rather strongly that distance is relative (to the chosen rod). If we tried to double every distance in the universe, the length of the rod would be doubled too, and nothing actually observable would be changed. For reasons like this, physicists have a hunch that absolute scale should have no objective meaning. However, this is not confirmed by existing theories and experimental facts, which do suggest that distance is in a well-defined sense absolute. The hope remains that a physics completely without scale will be found. If so, Shape Space gives an idea of what the corresponding Platonia will be like. There is still a uniquely distinguished point in it – the central point in Figure 8. It takes the place of Alpha in Triangle Land. If you 'touched' the central point, the equilateral triangle would 'light up'. This is the most symmetrical configuration the three-body universe can have. Symmetry is beautiful in one way but bland in another. The boundaries of Shape Space are somewhat unappealing too, because they represent improper triangles that flatten into a line (mathematicians call such configurations, collinear in this case, degenerate). The vertices of Shape Space correspond to collinear configurations in which one particle is infinitely far from the other two. The interesting structures in this case lie between the bland centre and the degeneracy of the frontiers.

In modern theoretical cosmology, distance is absolute and the universe expands. For reasons that are not yet understood, it simultaneously becomes more richly structured. In a cosmology without both time and scale, this would correspond, in a realistic scale-free Platonia, to going from the bland centre to the more interestingly structured 'instants of time' situated between it and the frontiers. That is where the mist I introduced in Chapter 3 must collect most thickly – have the highest intensity – at time capsules structured so that they seem to record evolution from the symmetric centre. This would be a cosmology of pure structure, an appealing thought. The scalene and obtuse triangles that inhabit the 'favoured belt' in Shape Space remind me of the line in Gerard Manley Hopkins's poem 'Pied Beauty', in which he praises 'All things counter, original, spare, strange'. In such a scheme, the bland centre and degenerate frontiers still have a vital role to play in the scheme of things. This is because a kind of resonance between all the instants of time determines where the mist settles. Any acoustician will recognize the importance of the walls and centre of a building in determining its harmonies. Platonia, shown here as Shape Space, is a 'heavenly vault' in which the music of the spheres is played. However, I should emphasize that the more realistic Shape Spaces corresponding to universes with more than three particles are most definitely 'open ended' and should not be thought of as enclosed spaces, as might appear from the simple example of Triangle Shape Space. Platonia is not a claustrophobic vault but an 'echoing canyon' open to the sky. What is more, there is a sense in which its echo, heard at any point within it, is what we call the past.

In completing this box, I note it has a nice unplanned symbolism. Box 3 considers Shape Space, which is represented by the perfect (equilateral) triangle, which itself, through each of its points, represents all triangles, all of which are unities in the sense mentioned on p.18. Shape Space illustrates Giordano Bruno's *monas monadum*, the unity of the unities.

If Laplace's divinity contemplates a three-particle universe, its history will be a curve in Triangle Land, which, omitting the triangle size, we can show as a curve in Shape Space. The example of real Newtonian three-body gravitational interaction shown in Figure 9 brings out very clearly the main fact that we associate with time, that its instants come in a unique succession. This translates beautifully into the winding path. (You can see why I am so indebted to Dierck Liebscher, who crafted this diagram.) However, the other two important attributes of time, duration and direction, are not yet reflected in Figure 9. There are no marks along the curve to indicate how much time elapses between any two points on it. It is also impossible to say in which direction along the curve time increases.

What information must we give to determine a history of the kind encoded in the path in Figure 9? According to Laplace, simply the positions and velocities of the bodies at one instant (together with their masses). Poincaré remarked, however, that the positions and motions of the bodies are defined in absolute space, and the speeds of the bodies are defined using absolute time. This is a subtle and important qualification.

If only relative quantities count, then Newton assumed too much structure. In a universe of just three particles, only the three distances between them (the triangle they form) should count. The triangle universe cannot have an overall position and orientation in some invisible containing space. Similarly, since the idea of an external 'grandstand clock' is absurd, we cannot say 'how fast' the universe travels along the curve in Figure 9. It simply occupies all the points along it.

If Mach is right, so that time is nothing but change and all that really counts in the world is relative distances, there should be a perfect analogue of Laplace's scenario with a divine intelligence that contemplates Platonia. Machian dynamics in Platonia must be about the determination of paths in that timeless landscape. It should be possible to specify an initial point in Platonia and a direction at that point, and that should be sufficient to determine the entire path. Nothing less can satisfy a rational mind. The history in Figure 10 starts at the centre of Shape Space, so that there the particles form an equilateral triangle, and set off in a certain direction. In Machian dynamics, the initial position and initial direction (strictly in Triangle Land, not Shape Space) should determine the complete curve uniquely. Now we can test this idea in the real world. The heavens provide plenty of triple-star systems, and astronomers have been observing their behaviour for a long time. They certainly meet the

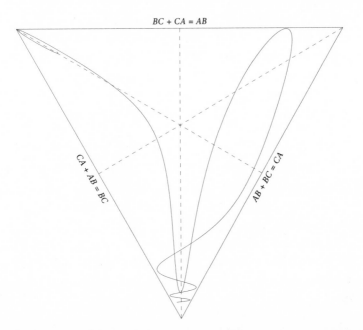

Figure 9 The (computer-generated) path traced in Shape Space by the triangles formed by three mutually gravitating particles. It is important to realize that the winding path shown is not traced by a single particle moving over the page, but that each point on the path represents the shape of a complete triangle. The curve shows the succession of triangle shapes. As explained in the text, it is impossible to say that time increases as you go in a particular direction along this curve. However, suppose we imagine it starts in the top left corner. This corresponds to particles A and C nearly colliding, while particle B is far away. Then they move to a configuration that is quite close to an equilateral triangle, after which A and B get very close together near the bottom of the diagram. Then the triangle shape evolves along the curve up to the top right. Where the curve nearly touches the top line, all three particles are almost on a line, with C between A and B. Finally the curve returns to the bottom of the figure, where the wiggles indicate that particles A and B are orbiting around each other, while C is far away. You see how history is all coded in one curve, but you just cannot tell in which direction it unfolds!

Laplace-type condition when described in Newtonian terms. But are their motions comprehensible from a Machian point of view? This is the question Poincaré posed.

APPARENT FAILURE

The answer is very curious. The motions are nearly but not quite comprehensible. This can be highlighted by showing how different possible

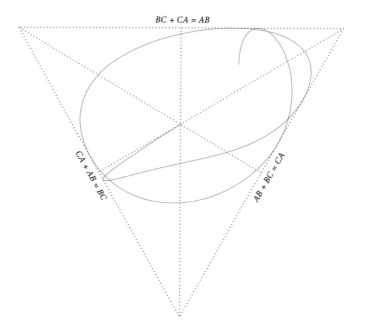

$$BC + CA = AB$$

CA + AB = BC

AB + BC = CA

Figure 10 Another possible path traced by the same three particles as in Figure 9 (I refer to them now as bodies). This history starts (or ends if time is assumed to run the other way) at the configuration in which the three bodies form an equilateral triangle. The shape of the triangle changes in a definite way at all points along the curve. If it left the initial equilateral triangle along one of the dotted lines, that would mean two sides of the triangle remaining equal in length while the third changes – the equilateral triangle would become an isosceles triangle. In fact, for the example shown, the ratios of all three sides change. For readers used to thinking of motions in ordinary space, this example corresponds to particles that constantly orbit each other in a fixed plane. The positions at which the curve touches the dotted line that bound Shape Space correspond to eclipses, when one particle is between the other two and on the line joining them. Such a configuration is called a *syzygy* (that's a nice word to show off with). In ordinary space, the one particle passes through the line joining the other two and comes out the other side. But the points on the curve in Figures 9 and 10 stand for the complete triangle, not one of the three particles. This is why the curve approaches the syzygy frontier and then returns into the interior of Shape Space. There are no triangles outside the syzygies!

Newtonian motions look when represented as curves in Triangle Land, our model Platonia – or rather Shape Space, since this is much easier to represent. To create a vivid picture, let us imagine that we are holding two cardboard triangles that are slightly different. These can represent the relative configurations of three mutually gravitating bodies at two slightly different instants of Newton's absolute time.

Playing the role of Laplace's divinity, we place the first triangle, at the instant when Newton's grandstand clock says it is noon, at some position in absolute space. A second later, we place the second triangle somewhere near it in a slightly different position. The first triangle defines the initial positions of the three bodies. Given the position of the second triangle one second later, we can calculate the initial motions, since we know where the particles have gone and how long it took them. (Strictly, to calculate the instantaneous velocities we must take an infinitesimal time interval, not one second, but that is a minor detail). Imagine now that a strobe light illuminates the bodies with a flash once every second, corresponding to the seconds ticking on Newton's clock, so that we can watch how the triangle formed by the bodies moves through absolute space. We have seen this already, in Figure 1. We can also plot the points corresponding to the triangles in either Triangle Land or Shape Space, obtaining a curve like those in Figures 9 and 10. This abstracts away the extra Newtonian information – the positions in absolute space and the time separations – that we possessed originally.

Now, wherever we place the two triangles, the resulting curves in either Triangle Land or Shape Space will all start at the same point, since we always begin with the same triangle, and that corresponds to just one fixed point in Platonia. The curve must also have the same initial direction, since that is determined by the position of the second triangle in Platonia, which is also fixed. This is explained in the caption to Figure 10. The question is, how does the curve run after that? What effect do the positioning of the first two triangles in absolute space and the time separation have on the subsequent evolution?

To answer this question, we need the notion of centre of mass (Box 4). For a given triangle, there are two different things to bear in mind when it comes to placing it in absolute space. First, we can place its centre of mass anywhere. Since space has three dimensions, this means that we can shift the centre of mass along three different directions. Physicists say that in such cases there are three *degrees of freedom*. Second, holding the centre of mass fixed, we can change the orientation of the triangle in space. This introduces three more degrees of freedom. To see this, picture an arrow passing through the centre of mass perpendicular to the triangle. It will point to somewhere on the two-dimensional sky, giving two degrees of freedom. The third arises because one can, keeping the arrow fixed, rotate the triangle around it as an axis.

BOX 4   Centre of Mass

The *centre of mass* of a system of bodies is the position of a fictitious mass equal to the sum of the masses of the system. For two unequal masses m and M, it lies on the line joining them at the position that divides the line in the ratio M/m – that is, closer to the heavier mass in that proportion. For any isolated system of bodies, the centre of mass either remains at rest or moves uniformly in a straight line through absolute space. The centre of mass for three bodies is shown in Figure 11.

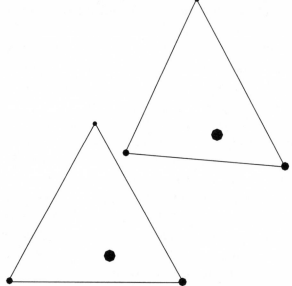

Figure 11 Masses of 1, 2 and 3 units (indicated by their sizes) are shown at the vertices of the two slightly different triangles (which correspond to the two triangles discussed in the main text). The two centres of mass are shown by the big blob (mass 6 units). The position of the centre of mass is found by finding the centre of mass of any pair and then the centre of mass of it and the remaining third mass.

Now, wherever we place the centre of mass of the first triangle, and however we orient it, the sequence of triangles that then arises is always the same. The path traced out in Platonia is the same. The starting position in absolute space does not matter an iota. This is rather remarkable. It is as if you could grow identical carrots in your garden, at the bottom of the sea, and in outer space. Different locations in absolute space have a decidedly shadowy reality. Unlike real locations on the Earth, they do not have any observable effects.

The 'location in time' is equally difficult to pin down. We started our experiment at noon according to Newton's clock. In fact, the starting time has no influence whatsoever: all the 'carrots' come out just the same. So, as far as the position of the first triangle in both absolute space and time is concerned, it has no influence whatsoever. We begin to wonder whether they play any role at all. This doubt is strengthened when we consider where to place the second triangle. It turns out that we can position its centre of mass anywhere in absolute space relative to the first triangle. This too has no effect at all on the sequence of triangles that then follow. This absence of effect is due to so-called Galilean relativity, which is one of the most fundamental principles of physics (Box 5).

BOX 5   The Galilean Relativity Principle

Galileo noted that all physical effects in the closed cabin of a ship sailing at uniform speed on a calm sea unfold in exactly the same way as in a ship at rest. Unless you look out of the porthole, you cannot tell whether the ship is moving. Quite generally, in Newtonian mechanics the uniform motion of an isolated system has no effect on the processes that take place within it. The left-hand diagram in Figure 12 shows (in perspective) the triangles formed by the three gravitating bodies in the history of Figure 10 at equal intervals of absolute time. The individual bodies move along the 'spaghetti' tubes. The centre of mass moves uniformly up the z axis. (Despite appearances, the triangles are always horizontal, i.e. parallel to the xy plane.) The right-hand diagram has two physically equivalent interpretations. First, it is how observers moving uniformly to the left past the system on the left would see that system receding behind them. Second, it is also how observers at rest relative to the system on the left would see a system identical to that system except for a uniform motion of the centre of mass to the right. This is how the happenings in the cabin of Galileo's galley would be 'sheared' to the right for observers standing on the shore. Depending on the speed of the system, the centre of mass will be shifted in unit time by different amounts, but the actual sequence of triangles remains the same. This corresponds to the freedom mentioned in the text.

Because of the relativity principle, the laws of motion satisfied by bodies take exactly the same form in any frame of reference moving uniformly through absolute space as they do in absolute space itself. Although Newton did not like to admit it, this fact makes it impossible to say whether any such frame, which is called an *inertial frame of reference*, is at rest in absolute space or moves through it with some uniform velocity. Bodies with no forces acting on them move in a straight line with uniform speed in any inertial frame of reference (hence the name). It is impossible to say that you are at rest in absolute space, only that you are at rest in some inertial frame of reference. For historical reasons I use 'absolute space' in the text, but strictly I should be using 'any inertial frame of reference'.

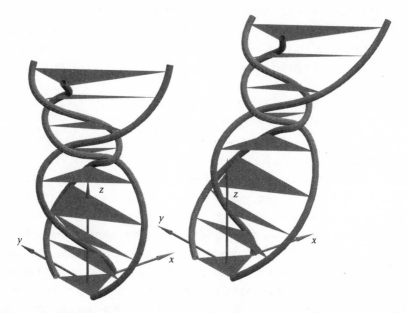

Figure 12 Unlike Figures 9 and 10 (and the later Figure 14), the lines followed by the spaghetti strands in this figure (and also Figure 13) show the tracks of the three individual particles in space. This is why there are three strands and not a single curve. It will help you a lot if you can get used to thinking about these two different ways of representing one and the same state of affairs. Here we see individual particles moving in absolute space. In Figures 9, 10 and 14 we 'see' (in our mind's eye) the 'world' or 'universe' formed by the three particles moving in Platonia.

There are only four freedoms that remain. Having placed the centre of mass of the second triangle at some position, we can change its orientation (three freedoms). We can also change the amount of Newton's absolute time that elapses between the instants at which the three bodies occupy the two positions (one freedom, the fourth). If the time difference is shortened, this means that the bodies travel farther in less of Newton's time – that is, they are moving faster initially. In fact, since the motion of the centre of mass does not matter, we can keep it fixed and change only the orientation. Now, at last, we come to something that does matter. Both these changes – in the time difference and in the relative orientation – have dramatic consequences, which are illustrated in Figures 13 and 14.

Figures 13 and 14 express the entire mystery of absolute space and time. Both of Newton's absolutes are invisible, yet their effects show up in the evolutions of the triangles, which are more or less directly visible. The astronomers *do* see stars and the spaces between them (admittedly in projection) when they look through telescopes. If time were merely change and only distances had dynamical effect, a decent Machian mechanics – one that would satisfy Laplace's divine intelligence – should lead to exactly the same evolutions in all nine cases. This is manifestly not true for the real triple-star systems that astronomers observe. All the different kinds of evolution shown in Figures 13 and 14, and many more, are found. All the facts that enabled Newton to win his argument against Leibniz are contained in these diagrams, but it took about two centuries before Poincaré found the best way to demonstrate them. He concluded regretfully that a mechanics that uses only relative quantities, as Mach advocated, cannot get off the ground. It lacks perfect Laplacian determinism. Nevertheless, the failure is curious. Absolute space and time could have had an effect through all the freedoms allowed in the placing of the two triangles. There are fourteen degrees of freedom in total, of which ten have no effect whatsoever. This is just what the invisibility of space and time would lead us to expect. Yet four degrees of freedom do have a profound influence. Three are associated with twists in space, the fourth with the overall speed put into the system. These strange mismatches between expectation and reality have kept the philosophers arguing and the physicists puzzling for centuries.

The fact is that Newton's absolute space and time play a decidedly odd role. The first problem is their invisibility. The more serious problem is what little part they play in the whole story, and how irrationally they enter the stage when they do participate in the action. Once we have chosen the relative orientation and time separation of the two triangles, we can take them anywhere in absolute space and time. They will always give rise to the same evolution. Absolute space and time seem to matter very little; only the relative orientation and time separation count.

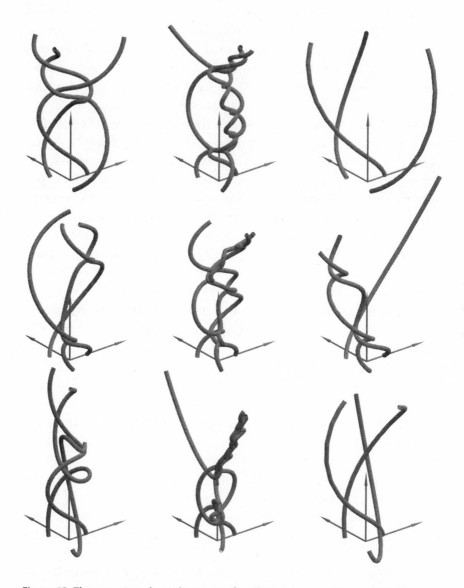

Figure 13 These are 'spaghetti diagrams' of evolutions in absolute space like the left-hand one in Figure 12 (the one at the top left is the same evolution but with the triangles removed). The corresponding curves in Shape Space are shown in Figure 14. In each diagram the evolution commences with the three bodies forming an equilateral triangle, and all the corresponding curves start in the same direction in Triangle Land and Shape Space. This is because the second triangle is the same in both cases. The different evolutions are created by giving the bodies different initial speeds (they are different in the three rows) and by giving the triangles different orientational twists (different in the three columns).

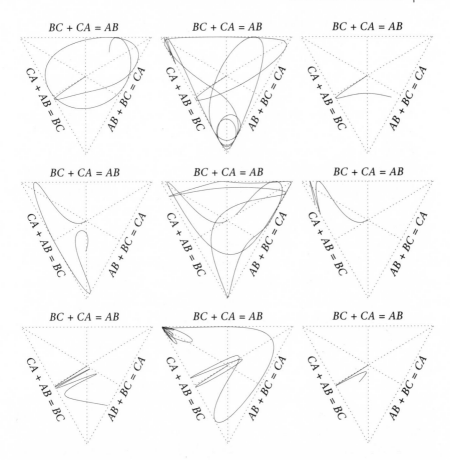

Figure 14 These are the curves in Shape Space corresponding to the nine evolutions in absolute space shown in Figure 13. They all start from the same point with the same direction, but then diverge strongly. Remember, as I explained in the caption to Figure 9, that these curves represent not the motion of a single particle across the page, but the shapes of continuous sequences of triangles. If you 'stuck a pin' into any point on one of these curves, the triangle corresponding to it would 'light up'. It is very important to appreciate that Figures 13 and 14 show identical happenings in two different ways. Since Newton's time, nearly all physicists have believed the Newtonian representation, Figure 13, to be the physically correct way to think about these things.

Following Leibniz and Mach, I believe Figure 14 is the right way. However, this approach faces a severe difficulty explained in the text. It is only in Chapter 7 that I shall explain how it is overcome.

But these are our arbitrary choices. Once we have chosen two triangles, nothing about the triangles in themselves gives any hint as to how we should make the choices. Leibniz formulated two great principles of philosophy that most scientists would adhere to. The first is the *identity of indiscernibles*: if two things are identical in *all* their attributes, then they are actually one. They are the same thing. The second, which we have already met, is the principle of *sufficient reason*: every effect must have a cause. There must be some real observable difference that explains different outcomes.

Now we can see the problem. Considered in itself, the pair of triangles is just one thing. Each different relative orientation and time separation we give them depends on our whim. They should not have any effect. Yet each has momentous consequences: they create quite different universes. An exactly analogous problem arises if the universe consists of any number of particles. Two snapshots (the analogues of the two triangles) of the relative configuration of the universe are never quite enough to determine an entire history uniquely.

Before we look at the one possibility that can resolve this puzzle, it is worth considering how the four freedoms that do count show up in practice. We shall then be able to see what a great discovery Newton's invisible framework was. We start with the twists.

SPACE AND SPIN

When I was a boy, there was only one sport at which I was any good: the high jump. One year I went on a training course at the athletics ground in Oxford. We were introduced to *angular momentum* and how it could be exploited to improve the jump. As tallest of the young hopefuls, I was chosen to give a demonstration. The instructor made me lie on my side, arms and legs outstretched, on a small bench turntable. He started to rotate it slowly and asked, 'If you pull yourself into a crouched position, what will happen?' I knew, I was studying physics: 'Angular momentum will be conserved, and the turntable will spin faster.' 'Right,' he said, 'do it.' Proudly, I pulled in my arms and legs with vigour. The effect was frightening. The turntable whizzed around so fast that I panicked, tried to get off, and was thrown onto the floor. I escaped with bruises. I am still kicking myself, not about the accident, but because I did not stay on another day. I would have seen Roger Bannister run the first four-minute mile.

Angular momentum is a kind of net spin about a fixed axis. To calculate it for the Earth, you multiply the mass of each piece of matter in the Earth by its perpendicular distance from the rotation axis and the speed of its circular motion about the axis. The Earth's total angular momentum is the sum of the contributions of all the pieces. Clockwise and anticlockwise motions count oppositely. A jet plane flying round the world in the opposite sense to the daily rotation contributes with the opposite sign.

By Newton's laws, this net spin cannot change for an isolated system. This universal law applies equally to humans and planets. When I pulled in my arms and legs in Oxford, I abruptly reduced the distance of much of my mass from the rotation axis. This inescapably enforced an equally abrupt increase of my rotational speed – with its unfortunate consequences. The same law explains why the Earth's rotation axis stays fixed, pointing towards the pole star, and why the length of the day, the rotation period, does not change. The rotation speed could change only if the Earth could expand or contract, but, being rigid, it cannot. (Actually, both the axis and the day do change very slowly due to the external influence of the Sun and Moon.) For rigid bodies like the Earth and a top, the effects of angular momentum are rather obvious. However, its effects are far-reaching.

A globular cluster may contain a million stars. It has no rigidity – all its stars move individually in different directions, though gravity holds the cluster together. Its angular momentum is found by choosing three mutually perpendicular axes, and calculating the net spin around each of them. These correspond exactly to the three degrees of freedom to make twists, mentioned in the previous section. However, the three axes can always be chosen in such a way that the spin about two of them is zero, and all the net spin is thus about a single axis. This axis is a kind of arrow that points in a certain direction in space. It and the net spin remain completely unchanged as time passes. In astronomy, time passes in aeons. Since the stars all move in different directions, the bookkeeping exercise that nature performs is remarkable. A deep principle is at work.

The laws of nature are seldom seen to be operating in a pure form, and are hard to recognize. Air resistance and friction distort the basic laws of mechanics. But the greatest difficulty arises because the laws involve time, and we experience only one instant at a time. If only we could see all the instants of time stretched out before us, we could see the effects of the laws of motion directly, as in some of the diagrams earlier in the book.

However, a few phenomena reveal mechanics at work in a striking fashion. They are often associated with angular momentum. The humble top is one of the best examples. Riding a bicycle is another: the reassuring way in which balance is maintained as you speed down a hill with the air rushing past you is down to the angular momentum in the spinning wheels. Once the wheels are turning fast, they have a strong tendency to keep their axis of rotation horizontal. Indeed, a child's hoop illustrates beautifully how the rotation axis maintains a fixed direction. So does the frisbee, spinning true as it floats through the air. Much grander examples occur naturally. I have already mentioned the earth's rotation, which we see as the rising and setting of the Sun, Moon and stars and their ceaseless march across the sky. Many of our images of time come from this phenomenon, the child's top writ large.

However, in all these examples there is a rigid body. The example of globular clusters tells of a mighty invisible framework behind the all too elusive phenomena. Newton knew it was there long before the astronomers found the grandest examples of its handiwork: spiral galaxies. In them, the initially invisible effects of the framework have become visible. Indeed, any isolated collection of matter, whatever its nature – a million stars in a globular cluster or a huge cloud of dust in space – has its associated fixed axis of net spin. Laplace called the plane perpendicular to it through the centre of mass the *invariable plane*, because its orientation can never change. Sometimes it can actually be seen. This is because some motions can be changed or even lost through mutual interactions, whereas others cannot. For example, objects moving parallel to the spin axis in opposite directions may collide and be deflected into the invariable plane. Over time, the matter in the system can 'collect' in or near it provided there is still the correct amount of circular motion about the axis. This has happened in spiral galaxies, in which the bright stars in the spiral arms are formed from such accumulated matter. They 'light up' the invariable plane, making it visible (Figure 15).

A similar effect has been at work in the solar system. About four and a half billion years ago the Sun and planets formed from a huge cloud of dust left over from a supernova explosion. The dust had some net spin, and an associated invariable plane. The Sun formed near the cloud's centre of mass, and gathered up most of the mass in the cloud. More or less all of the solar system's rotation now takes place in the plane of the ecliptic, in which the Earth moves around the Sun. Although the Sun got the bulk of the mass, Jupiter has most of the spin.

Figure 15 A spectacular spiral galaxy seen 'from above'.

The fact that all the planets move in the same direction around the Sun in nearly coincident planes is thus a remote consequence of the relatively modest initial net spin of the primordial dust cloud. We see the result in the sky, since all the celestial wanderers – the Sun, Moon and planets – follow much the same track against the background of the stars. Ironically, Newton underestimated the power of his own laws. He could not bring himself to believe that the solar system had arisen naturally. 'Mere mechanical causes', he said, 'could not give birth to so many regular motions.' He asserted that 'this most beautiful system' could only have proceeded 'from the counsel and dominion of an intelligent and powerful Being'. One wonders what Newton would have made of the modern pictures of Saturn and its rings (Figure 16). Of all the images created in the heavens by gravity and the invariable plane, this is surely the most perfect.

For three centuries, the best explanation for phenomena like the rings of Saturn has remained Newton's: inertia, the inherent tendency of all objects to follow straight lines in the room-like arena of absolute space. If these are accepted, then the rings of Saturn, tops, frisbees and all the other manifestations of angular momentum can be explained. However, Newton's account is not so much an explanation as a statement of facts in need of explanation. Since it is always matter that we actually see, should we not try to account for these things without the mysterious

Figure 16 Saturn and its rings.

intermediaries of absolute space and time? Before we attack this problem, we need to consider energy and, in the next chapter, clocks and the measurement of time.

ENERGY

Energy is the most basic quantity in physics. It comes in two forms: kinetic energy measures the amount of motion in a system, while potential energy is determined by its instantaneous configuration. Like angular momentum, in an isolated system the sum of the two remains constant. If one decreases, the other must increase. For example, the potential energy of a falling body is proportional to its height and decreases as it falls. The speed of descent, and with it the kinetic energy, increases by an exactly compensating amount.

Energy, like the whole of mechanics, has a curious hybrid nature. Absolute space and time are needed to calculate kinetic but not potential energy. Each body of mass m and speed v in a system contributes a kinetic energy $\frac{1}{2}mv^2$. The speed is measured in absolute space, which is why it is needed to calculate kinetic energy. By contrast, the potential energy of a system depends only on its relative configuration. For example, each pair of gravitating bodies in a system contributes to the system's total

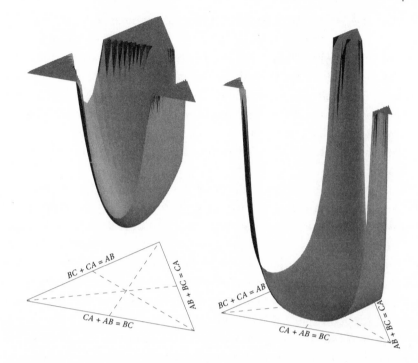

Figure 17 The gravitational potential energy of three bodies of different masses is shown as the height of a surface above Shape Space (Figure 8), each point on which corresponds to a different shape of the triangle formed by the three bodies. The overall scales of the configurations on the right are nine times greater, so the magnitude of the potential energy is much lower. Since potential energy is inversely proportional to separation, it increases sharply towards the corners of Shape Space, corresponding to two-particle coincidences, and becomes infinite at them. As this cannot be shown in the figure, the surfaces have been cut off at a certain height. The most distant corner of Shape Space corresponds in this figure to coincidence of the two most massive particles, so this is why the potential increases most strongly there.

potential energy an amount that is inversely proportional to their separation. If this is doubled, the potential energy of the pair is halved. Since each point in any Platonia corresponds to a different configuration of bodies, the potential energy changes from place to place in Platonia. This is illustrated for three bodies in Figure 17.

Like angular momentum, the energy affects the appearance of systems and the behaviour of individual objects. For gravity the potential energy is negative, while the kinetic energy is positive. Thus the total energy E can be either positive, zero or negative. If a spacecraft is launched with sufficient speed, it can escape from the Earth's gravity because its E is positive. If E is zero, the spacecraft has exactly the escape velocity, and escape is just possible. If E is negative, the spacecraft cannot escape from the Earth and will either orbit the Earth or fall back to ground. The planets can never escape from the Sun because they have negative E. Star clusters can remain concentrated in a relatively small region of space only if their energy is negative, otherwise they would rapidly disperse. This is why we do see such fine objects as the galaxies and star clusters in the sky. It is also largely the reason why the Sun and planets have their beautiful round shapes.

Thus, the shapes of almost all the objects astronomers observe in the sky reflect their energy and angular momentum. They, in turn, seem impossible to explain unless absolute space and time do exist and have a real influence, just as Newton claimed. The evidence for Newton's invisible framework is written all over the sky. The evidence can be summarized as the *two-snapshots problem*. Suppose that snapshots of an isolated system taken at two closely spaced instants show only the separations of its bodies, not the overall orientations in absolute space. The separation in time between the snapshots is also unknown. If the system is a globular cluster, the snapshots contain millions of data. However, to determine the evolution of the system, four pieces of data are still lacking. They determine the kinetic energy (one piece of data) and the angular momentum (three pieces of data). Although they cannot be deduced from the two snapshots, they have a huge influence on the evolution, which can often be seen at a glance. A third snapshot will yield the data, but also much redundant information. The four missing pieces of data comprise the entire evidence for absolute space and time. Every system in the universe proclaims their existence. This seems to make nonsense of my claim that time does not exist. There appears to be more to the universe than its relative configurations. There is invisible structure, of which no trace can be found in Platonia.

CHAPTER 6

The Two Great Clocks in the Sky

WHERE IS TIME?

Newton's mysterious 'timepiece' and speeds measured relative to it figured prominently in the last chapter. But is it really there, and how can we ever read its time if it is invisible? This chapter is about these two questions.

A simple but famous experiment of Galileo provides strong evidence for something very like Newton's absolute time. He rolled a ball across a table and off its edge. His analysis of its fall was a major step in mechanics. First he noted the ball's innate tendency to carry on forward in the direction it had followed on the table. It also started to fall under gravity, picking up speed. Galileo conjectured that two processes were at work independently, and that each could be analysed separately. The total effect would be found by simply adding the two processes together.

Galileo's recognition of the tendency to keep moving forward anticipated Newton's law of inertia. He did not recognize it as a universal law, but he did make it precise in some special cases. For the example of the ball, he conjectured that but for gravity (and air resistance) the ball would move for ever forward with uniform speed. (He actually thought that the motion would be around the Earth – Galileo's inertia was circular. Luckily, the difference was far too small to affect his analysis.)

As for the second process, Galileo had already found that if an object is dropped from rest and in the first unit of time falls one unit of distance, then in the next it will fall a further three, in the next five, and so on. He was entranced by this, and called it the *odd numbers rule*. Now consider the sequence:

at $t = 1$, distance fallen = 1,
at $t = 2$, distance fallen = 1 + 3 = 4,
at $t = 3$, distance fallen = 1 + 3 + 5 = 9,
at $t = 4$, distance fallen = 1 + 3 + 5 + 7 = 16,...

The distance fallen increases as the square of the time: $1^2 = 1$, $2^2 = 4$, $3^2 = 9$, $4^2 = 16$,.... Galileo's originality was to seek for a deeper meaning in this pattern.

Many teenagers can now do in seconds a calculation that took Galileo a year or more – it was so novel. He asked: if the distance fallen increases as the square of the time, how does the speed increase with time? He eventually found that it must increase uniformly with time. If after the first unit of time the object has acquired a certain speed, then after the second it will have twice that speed, after the third three times, and so on. Galileo's work showed that, in the absence of air resistance, a falling body always has a constant acceleration. It never ceases to amaze me what consequences flowed from Galileo's simple but precise question. It taught his successors how to read the 'great book of nature' (Galileo's expression). From a striking empirical pattern, he had found his way to a simpler and deeper law.

To analyse the falling ball, Galileo simply combined the two processes – inertia and falling – under the assumption that each acts independently. He obtained the famous parabolic motion (Figure 18). In each unit of time, the ball moves through the same horizontal distance, but in the vertical dimension the distance fallen grows as the square of the time. The resulting curve traced by the ball is part of a parabola. Newton applied Galileo's method for terrestrial motions to the heavens, and showed that the laws of motion had universal validity. This was the first great unification in physics. There may be a lesson for us here in our present quest – the search for time. We may have to look for it in the sky.

A search is needed. It is striking that all the elements in Galileo's analysis are readily visualized. You can easily call up a table and the parabola traced by the ball in your imagination. Yet one key player seems reluctant to appear on the stage. Where is time? This is the question I have so far dodged. It presents a severe challenge to the idea that configurations are all that exist. Suppose that we take snapshots of the ball as it rolls across Galileo's table in Padua, where he experimented. These snapshots can show everything in his studio. However, only the ball is moving. We take lots of snapshots, at random time intervals, until the ball is just about to fall over the edge. We put the snapshots, all mixed up, in a bag and, supposing time travel is possible, present it to Galileo and ask him whether, by examining the snapshots, he can tell where the ball will land.

He cannot. Had we rolled the ball twice as fast, it would have passed

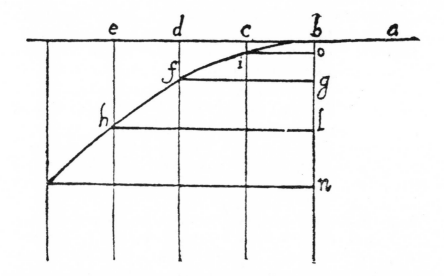

Figure 18 Galileo's own diagram of parabolic motion. The ball comes from the right and then starts to fall. Incidentally, this diagram illustrates how conventions get established and become rigid – a modern version of it would certainly show the ball coming from the left and falling off on the right.) The uniformity of the horizontal inertial motion is shown by the equality of the intervals *bc*, *cd*, *de*, …, while the odd-numbers rule is reflected in the increasing vertical descents *bo*, *og*, *gl*, … .

through the identical sequence of positions to the table's edge, and they are all the snapshots capture. The speed is not recorded. But the ball's speed determines the shape of the parabola, and hence where the ball lands. In fact Galileo will not even know in which direction the ball is going. Perhaps it will fall off the right-hand side of the table. More clearly than with the three-body evolutions, which mix the effects of time and spin, we see here the entire evidence for absolute time. The speed determines the shape of the parabola. There is manifestly more to the world than the snapshots reveal. What and where is it? Galileo himself provides an answer of sorts. He tells us that he measured time by a water-clock – a large water tank with a small hole in the bottom. His assistant would remove a finger from the hole and let the water flow into a measuring flask until the timed interval ended. The amount of water measured the time.

We have only to include the water tank and assistant in the snapshots, and everything is changed. Galileo can tell us where the ball will land because he can now deduce its speed. There are some important lessons we can learn from this. First, it is water, not time, that flows. Speed is not distance divided by time but distance divided by some real change elsewhere in the world. What we call time will never be understood unless this fact is grasped. Second, we must ask what change is allowed as a measure of time. Galileo measured the water carefully and made sure that it escaped steadily from the tank – otherwise his measure of time would surely have been useless. But the innocent word 'steadily' itself presupposes a measure of time. Where does that come from? It looks as if we can get into an unending search all too easily. No sooner do we present some measure that is supposed to be uniform than we are challenged to prove that it is uniform.

It is an indication of how slowly basic issues are resolved – and how easily they are put aside – that Newton highlighted the issue of the ultimate source of time nearly two hundred years before serious attempts were made to find it. Even then, the attempts remained rather rudimentary and few scientists became aware of them. It is interesting that Galileo had already anticipated the first useful attempt. This was actually forced upon him by the brevity of free fall in the ball experiment: it was all over much too quickly for the water-clock to be of any use. (It came into its own when Galileo rolled balls down very gentle inclines.) To analyse the parabola, he found a handy substitute. He noted that if the horizontal motion of the falling ball does persist unchanged, then the horizontal distance traversed becomes a direct measure of time. He therefore used the horizontal motion as a clock to time the vertical motion. His famous law of free fall was then coded in the shape of the parabola. Its defining property is that the distance down from the apex (where the ball falls off the table) increases as the square of the horizontal distance from the axis. But this measures time.

Thus, time is hidden in the picture. The horizontal distance measures time. It would be nice if one could say 'the horizontal distance *is* time'. This is the goal I am working towards: time will become a distance through which things have moved. Then we shall truly see time as it flows, because time will be seen for what it is – the change of things. However, there are many different motions in the universe. Are they all equally suitable for measuring time? A second question is this: what causal connections are at work here? Galileo measured time by the flow of water, but it is hard to believe that a little water flowing out of a tank in the corner of his studio directly caused the balls to trace those beautiful parabolas through the air. If time derives from motion and change – and it is quite certain that all time measures do – what motion or change, in the last resort, is telling the ball which parabola to trace? The first question is more readily answered.

THE FIRST GREAT CLOCK

Nearly two thousand years ago, astronomers knew that some motions are better than others as measures of time. This they discovered experimentally. For the early astronomers, there were two obvious and, on the face of it, equally good candidates for telling time. Both were up in the sky and both had impressive credentials. The stars made the first clock, the Sun the second.

The stars remain fixed relative to each other and define *sidereal time*. Any star can be chosen as the 'hand' of the stellar clock: one merely has to note when it is due south. The stellar clock then ticks whenever that star is due south (i.e. when it crosses the meridian). Fractions of the 'tick unit' are measured by its distance from the meridian. A mere glance at the night sky could tell the ancient astronomers the time to within a quarter of an hour. With some care, times could be told to a minute. There is something wonderful about this great clock in the sky. It was a unique gift to the astronomers. The discoveries that culminated with Kepler's laws of planetary motion, and many more made until well into the twentieth century, are unthinkable without it. No other phenomenon in nature could match it for convenience and accuracy. In millennia it has lost a few hours.

But there is a rival – the Sun. It defines *solar time*. This is the clock by which humanity and all other animals have always lived. The principle is the same: it is noon when the sun crosses the meridian. You don't even have to be an astronomer to tell the time by this clock; a sundial will do.

Merely describing the clocks shows that speed is not distance divided by time, but distance divided by some other real change, most conveniently another distance. Roger Bannister ran one mile in four minutes; normal mortals can usually walk four miles in one hour. What does that mean? It means that as you or I walk four miles, the sun moves 15° across the sky. But this is not quite the complete story of speed and time, because there is a subtle difference between the two clocks in the sky –

they do not march in perfect step. One and the same motion will have a different speed depending on which clock is used. One difference between the clocks is trivial: the solar day is longer than the sidereal. The Sun, tracking eastwards round the ecliptic, takes on average four minutes longer to return to the meridian than the stars do. This difference, being constant, is no problem. However, there are also two variable differences (Box 6).

BOX 6   The Equation of Time

The first difference between sidereal and solar time arises from one of the three laws discovered by Kepler that describe the motion of the planets. The Sun's apparent motion round the ecliptic is, of course, the reflection of the Earth's motion. But, as Kepler demonstrated with his second law, that motion is not uniform. For this reason, the Sun's daily eastward motion varies slightly during the year from its average. The differences build up to about ten minutes at some times of the year.

The second difference arises because the ecliptic is north of the celestial equator in the (northern) summer and south in the winter. The Sun's motion is nearly uniform round the ecliptic. However, it is purely eastward in high summer and deep winter, but between, especially near the equinoxes, there is a north–south component and the eastward motion is slower. This can lead to an accumulated difference of up to seven minutes.

The effects peak at different times, and the net effect is represented by an asymmetric curve called *the equation of time* (it 'equalizes' the times). In November the Sun is ahead of the stars by 16 minutes, but three months later it lags by 14 minutes. This is why the evenings get dark rather early in November, but get light equally early in January. The stars, not the sun, set civil time.

Since the Sun is much more important for most human affairs than the stars, how did the astronomers persuade governments to rule by the stars? What makes the one clock better than the other? The first answer came from the Moon and eclipses. Astronomers have always used eclipse prediction to impress governments. By around 140 BC, Hipparchus, the first great Greek astronomer, had already devised a very respectable theory of the Moon's motion, and could predict eclipses quite well.

Now, in the timing and predicting of eclipses, half an hour makes a difference. They can occur only when the Moon crosses the ecliptic – hence the name – and the Moon moves through its own diameter in an hour. There is not much margin for error. By about AD 150, when Claudius Ptolemy wrote the *Almagest*, it was clear that eclipses came out right if sidereal, not solar time was used. No simple harmonious theory of the Moon's motion could be devised using the Sun as a clock. But the stars did the trick.

What Hipparchus and Ptolemy took to be rotation of the stars we now recognize as rotation of the Earth. It is strikingly correlated with the Moon's motion. Even more striking is the correlation established by Kepler's second law, according to which a line from the Sun to a planet sweeps out equal areas in equal intervals of sidereal time. Whenever astronomers and physicists look carefully, they find correlations between motions. Some are simple and direct, as between the water running out of Galileo's water-clock and the horizontal distance in his parabolas; others, especially those found by the astronomers, are not nearly so transparent. But all are remarkable.

If two things are invariably correlated, it is natural to assume that one is the cause of the other or both have a common cause. It is inconceivable, as I said, that water running from a tank in Padua can *cause* inertial motion of balls in northern Italy. It is just as inconceivable that the spinning Earth causes the planets to satisfy Kepler's second law. Kepler, in fact, thought that it arose because all the planets were driven in their orbits by a spinning Sun, but we must look further now for a common cause. We shall find it in a second great clock in the sky. This will be the ultimate clock. The first step to it is the inertial clock.

THE INERTIAL CLOCK

The German mathematician Carl Neumann took this first step to a proper theory of time in 1870. He asked how one could make sense of Newton's claim, expressed in the law of inertia, that a body free of all disturbances would continue at rest or in straight uniform motion for ever. He concluded that for a single body by itself such a statement could have no meaning. In particular, even if it could be established that the body was moving in a straight line, uniformity without some comparison was meaningless. It would then be necessary to consider at least two bodies. He introduced the idea of an *inertial clock*. He supposed that one body was known to be free of forces, so that equal intervals of its motion could then be taken to define equal intervals of time. With this definition, it would be possible to see if the other body, also known to be free of forces, moved uniformly. If so, then in this sense Newton's first law would be verified.

Neumann's idea illustrates the truth that time is told by matter – something has to move if we are to speak of time. Unfortunately, it left unanswered at least three important questions. How can we say that a body is moving in a straight line? How can we tell that it is not subject to forces? How are we to tell time if we cannot find any bodies free of forces?

The answers to these questions will tell us the meaning of *duration*. If we leave aside for the moment issues related to Einstein's relativity theories and quantum mechanics, time as we experience it has two essential properties: its instants come in a linear sequence, and there does seem to be a length of time, or duration. I have tried to capture the first property by means of model instants. A random collection of such model instants would correspond to points scattered over Platonia. They would not lie on a single curve, and the fact that they do is, if verified, an experimental fact of the utmost importance. It enables us to talk about history.

But what enables us to talk so confidently of seconds, minutes, hours? What justification is there for saying that a minute today has the same length as a minute tomorrow? What do astronomers mean when they say the universe began fifteen billion years ago? Conditions soon after the Big Bang were utterly unlike the conditions we experience now. How can hours then be compared with hours now? To answer this question, I shall first assume that there are no forces in the world and that the only kind of motion is inertial. This simplification already enables us to get very close to the essence of time, duration and clocks. Then we shall see what forces do.

Suppose Newton claims that three particles, 1, 2 and 3, are moving purely inertially and that someone takes snapshots of them. These snapshots show the distances between the particles but nothing else (except for marks that identify the particles). We know neither the times at which the snapshots were taken nor any of the particles' positions in absolute space. How can we test Newton's assertion? We shall be handed a bag containing triangles and told to check whether they correspond to the inertial motion of three particles at the corners of the triangles. The Scottish mathematician Peter Tait solved this problem in 1883 (Box 7).

BOX 7   Tait's Inertial Clock

Tait used the relativity principle (Box 5) to simplify things. If the particles are moving inertially, one can always suppose that particle 1 is at rest; it is shown in Figure 19 as the black diamond where the three coordinate axes x, y, z in absolute space meet. Now, unless the particles collide at some time – and we need not bother about this exceptional case – there must come a time at which particle 2 passes 1 at some least distance a. The coordinate axes can be chosen so that the line of its motion is as shown by the string of black diamonds. We can choose the unit of time so that particle 2 has unit speed. It will be Neumann's clock, and each unit of distance it goes will mark one unit of time. Several positions of particle 2 are shown. They are the 'ticks' of the clock. At time $t = 0$, let particle 2 be at the point closest to particle 1 (at the black diamond on the x axis). At this time, particle 3 can be anywhere (three unknown coordinates) and have any velocity (three more unknowns). Thus, seven numbers are unknown: six for particle 3 and the distance a.

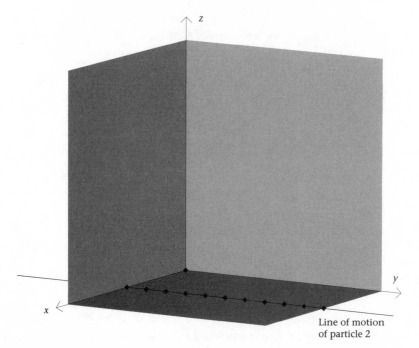

Figure 19 The arrangement of coordinates in Tait's problem.

Now, each snapshot contains three independent data – the three sides of the triangle at the instant of the snapshot. It would seem that three snapshots give nine data – more than enough for the accomplished Tait to solve the problem. But since we know none of the times at which the snapshots are taken, each gives only two useful data. Thus, four snapshots will give eight useful data, seven of which will establish the Newtonian frame into which the triangles fit, while the remaining one will verify that they are indeed obeying Newton's law. Figure 20 shows a typical solution.

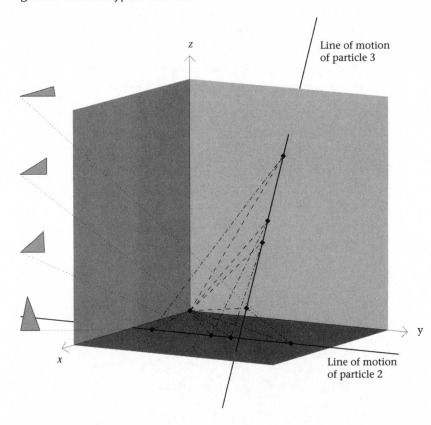

Figure 20 A solution of Tait's problem. The four 'snapshots' of the triangles (the given data) are shown in plan, with an indication of their positions in the 'sculpture' of four triangles created by the solution.

Tait's solution creates nothing less than Newton's invisible framework. It is both absolute space and the positions of the triangles in it at different instants of 'time'. We start with triangles. They are all that we have. We are told that they are not random triangles but have arisen through a law. We test this assertion and succeed in creating a simple 'sculpture' of three straight lines. The order created is dramatic. Apart from exceptional cases, the distances of the particles from one another do not change in a mutually uniform manner. In fact, they vary with respect to one another in a quite complicated way. Moreover, the triangles in themselves give no hint that there is any space in which their corners lie on straight lines. Yet such lines can be found.

It is equally remarkable that the motions along these lines are mutually uniform. Each particle is the 'hand' of a clock for the motion of the other two. The sculpture is a clock with many hands (two in this case, since particle 1 defines the origin). And, now that we have the rigid structure, we see that absolute space is redundant. The sculpture holds together on its own. The 'room' was never there until it was created from the triangles and rules. It is they that give an almost bodily tangibility to space and time.

They also explain the meaning of duration and the statement that a second today is the same as a second tomorrow. Duration is reduced to distance. If today or tomorrow any one of the 'hands' of the inertial clock moves through the same distance, then we can say that the 'same amount of time' has passed. The extra time dimension is redundant: everything we need to know about time can be read off from distances. But note how special is the distance that leads to a meaningful definition of duration. Any change of distance 'labels' the instants of time. In statements like 'Particle A hits B when C is five metres from D', 'five metres' identifies an instant of time – it labels the instant. That is sufficient for history. However, the obvious changes of distance – those between particles – do not lead to a sensible definition of duration. The secrets of time are rather well hidden.

A similar construction can be repeated for any number of particles – a hundred, a billion, quite enough to fill the sky and make a galaxy or even a universe. It is important that if there are more than five particles then three snapshots are already sufficient to solve Tait's problem, but two are never enough. This is very odd. For a thousand bodies, three snapshots contain far more information than we need, but two never give quite enough. The problem is exactly the one we encountered earlier. Two snapshots tell nothing about the relative orientations or the separation in time. We lack four pieces of information, and all the mystery of absolute space and time resides in them. We cannot make a clock until we get our hands on them. But when we have them, the properties that are revealed are very striking.

For example, Tait's construction is a good model for the motions of many thousands of stars that are relatively close neighbours of our Sun. They all 'fall' in the gravitational field of the Galaxy in much the same way. That motion hardly shows up in the relative separations. But also, because the stars are so far apart there is very little gravitational interaction between them. Their motions are thus well described by the construction. What is more, any three stars can be chosen to make a 'Tait clock' and tell the time. Any other three will make another. Thousands, millions of such clocks can be made. All these clocks, light years apart, keep time with one another.

I mentioned earlier the importance of not being misled by the special circumstances of our existence. One of them is the Earth. Only the tiniest fraction of the matter in the universe is in solid form. Indeed, only a small fraction of the Earth – its crust, on which we live, and the innermost core – is solid. This is our home, and we take it for the normal run of things. The ground, trees, buildings, hills and mountains make a framework, which is so like absolute space. It does seem quite natural that a body should move in a straight line in such a space. But we need to think what the universe in its totality is like. Take a billion particles and let them swarm in confusion – that is the reality of 'home' almost everywhere in the universe. The stars do seem to swarm, so do the atoms in the stars. To understand the real issues of timekeeping, we must imagine trying to do it in typical circumstances. We must master celestial timekeeping and not be content with the short cuts that can be taken on the Earth, for they hide the essence of the problem.

THE SECOND GREAT CLOCK

We have seen how to check whether bodies are moving inertially without prior access to absolute space and time. But all matter in the universe interacts. Interactions make things more complicated, and not only because the calculations are hard. If objects are moving inertially, any three will suffice to construct an inertial clock. But in a system of interacting bodies it is not possible to treat any of them separately because each is affected by the others. In addition, we can find no framework at all in which the bodies move uniformly in straight lines.

There are three parts to a clock: a mechanism, a clock-face and hands. The main problem of celestial timekeeping arises because the clock-face is invisible. A further problem is that the hands run at varying rates. Imagine an isolated system of three gravitationally interacting stars. We are again given only their relative positions, from which we are to construct a clock. Because of their interactions, no framework exists in which the three stars move along straight lines. The best we can achieve is some 'spaghetti sculpture' (Figure 13). This is found by telling a computer that there does exist a framework of absolute space and time in which the stars obey Newton's laws. However, the computer is given only the successive relative positions, not the positions in the framework at given times. But this is real information, and if the computer is given a sufficient number of snapshots it can search for an arrangement of them in a spaghetti sculpture in which the stars do obey Newton's laws. The positions in the framework and the separations in time are found by trial and error.

Suppose we are given ten snapshots. We can mark the positions of the triangles formed by the three bodies in Triangle Land (Figures 3 and 4). We can then tell the computer four of the positions. If the snapshots have indeed been generated by bodies that satisfy Newton's laws, the computer will find a curve that passes through them and the other six. We obtain a curve like those in Shape Space in Figures 9 and 10. We have to use both representations, in Triangle Land and in absolute space, because the raw data come to us in the former, but it is in the latter that we can make sense of them. Once we have solved the problem in absolute space, a timing of the evolution has been established. It is that timing of the events for which Newton's laws do hold. If the computer tried to assign other timings, they would not. The timing that does work can then be transferred back to the raw data: the curve in Triangle Land. We can make marks along it corresponding to the passage of the time found by the computer.

The same thing can be done for any number of bodies. Their relative configurations will correspond to different points along a curve in the corresponding Platonia. To lay out 'marks of equal intervals of time' on it, we have to go through the same procedure with the computer, telling it to find a framework and a time in which the bodies do satisfy Newton's laws. Only two facts about this process are significant. First, because all the bodies interact, all their positions must be used if the 'time marks' are to be found. To tell the time by such a clock, we need to know where all its bodies are. Time cannot be deduced from a small number of them, unlike inertial time; the clock has as many hands as the system has bodies. Second, no matter how many bodies there are in a system, the data in just two snapshots are never enough to find the spaghetti sculpture in absolute space and construct a clock. We always need at least some data from a third snapshot. As we have seen, this 'two-and-a-bit puzzle' is the main – indeed the only – evidence that absolute space and not Platonia is the arena of the universe.

You might think that this is all far removed from practical considerations. It is true that scientists have learned to make extremely accurate clocks using atomic phenomena. But this is a comparatively recent development. Before then, astronomers faced a tricky situation, which is worth recounting.

For millennia, the Earth's rotation provided a clock sufficiently reliable and accurate for all astronomical purposes. It was unique – the astronomers had access to no other comparable clock. However, about a hundred years ago, astronomical observations had become so accurate that deficiencies in it began to show up. Tidal forces of the Moon acting on the Earth sometimes give rise to unpredictable changes of the mass distribution in its interior. As my accident in Oxford demonstrated, such changes in a rotating body must change its rotation rate. The clock was beginning to fail the astronomers' growing needs for greater accuracy. Such crises highlight fundamental facts. What could the astronomers do?

They managed to find a natural clock more accurate than the Earth: the solar system. To make this into a clock, they assumed that Newton's laws governed it. (After the discovery of general relativity, small corrections had to be made to them, but this did not change the basic idea.) However, the astronomers had no direct access to any measure of time. Instead, they had to assume the existence of a time measure for which the laws were true. Making this assumption and using the laws, they could then deduce how all the dynamically significant bodies in the solar system should behave. Although they had no access to it, they then knew where the various bodies should be at different instants of the assumed time. Monitoring one body – the Moon, in fact – they could check when it reached positions predicted in the assumed time and verify that the other bodies in the solar system reached the positions predicted for them at the corresponding times. The astronomers were thus forced into the exercise just described, and they used the Moon as the hand of a clock formed by the solar system.

They originally called the time defined in this manner *Newtonian time*. It is now called *ephemeris time*. (An ephemeris is a publication which gives positions of celestial objects at given times.) For a decade or more it was actually the official time standard for civil and astronomical purposes. More recently, atomic time, which relies on quantum effects, has been adopted. There are several important things about ephemeris time. First, it is unthinkable without the laws that govern the solar system. Second, it is a property of the complete solar system (because all its bodies interact, all co-determine one another's positions). Third, it exists only because the solar system is well isolated as a dynamical system from the rest of the universe.

Ephemeris time may be called the unique simplifier. This is an important idea. If, as Mach argued, only configurations exist and there is no invisible substance of time, what is it that we call time? When we hold the configurations apart *in time* and put a duration between them, this something we put there is a kind of imagined space, a fourth dimension. The spacing is chosen so that the happenings of the world unfold in accordance with simple laws (Newton's or Einstein's). This is a consequence of the desire to represent things in space and time, and our inability hitherto to find laws of a simple form in any other framework.

Ephemeris time is the only standard we can use if clocks are to march in step. If we could not construct such clocks, we could never keep appointments and clocks would be useless. To see that there is only one sensible definition of duration, imagine that two teams of astronomers were sent to two similar but nevertheless different isolated three-body stellar systems. All they can do is observe their motions. From them they must generate time signals. Each team works separately, but the signals they generate must march in step – one clock may run faster than the other, but the relative rate must stay constant. There is only one measure of duration they can choose. In general, no motion in one system marches in step with any motion in the other. Only ephemeris time, deduced from the system as a whole, does the trick. A clock is any mechanical device constructed so that it marches in step with ephemeris time, the unique simplifier.

We can now see that there is only one ultimate clock: the universe.

Although it would not be practicable, if we wanted to obtain time of ultra-high accuracy from the solar system, we would, sooner or later, have to take into account the disturbances exerted by bodies farther away. Since there are no perfectly isolated systems within the universe, this process can only stop, if ever, when the entire universe has been made into a clock. The universe is its own clock.

In the light of this, let us think again about Galileo's ball rolling across the table in Padua. Snapshots of the ball alone were not sufficient to tell what would happen when it rolled over the edge. It seemed inconceivable that the ball's path could be determined by the little bit of water flowing from a tank used to tell the time. For such reasons as this, Newton rejected speed relative to any one motion as a fundamental concept and invoked instead speed relative to an abstract time. However, if we conceive the universe as a single dynamical entity, the abstract time becomes redundant. The speed of Galileo's ball that determines which parabola it will trace is its speed as measured by the totality of motions in the universe. This explains why some motions are distinguished from others for timekeeping. They are those that march in step with the cosmic clock, the unique true measure of time. This time is the distillation of all change. High noon is a configuration of the universe.

But the two-and-a-bit puzzle persists. We still have no simple direct way to measure time in Platonia, we always have to go through the intermediary of absolute space. This reflects the hybrid nature of energy. Kinetic energy is defined in absolute space, whereas potential energy is determined by instantaneous configurations and is thus independent of Newton's invisible framework. We shall be able to claim that Platonia is the arena of the world only if we can dispense with absolute space in the definition of kinetic energy. That is the next topic.

CHAPTER 7

Paths in Platonia

NATURE AND EXPLORATION

The two-and-a-bit puzzle is the statement that two snapshots of a dynamical system are nearly but not quite sufficient to predict its entire history. We need to know not only two snapshots, but also their separation in time and their relative orientation in absolute space. These are exactly the things that determine the energy and angular momentum of any system, both of which, as we have seen, have a profound influence on its behaviour.

There are two different ways to approach this problem. Either we assume the known laws of nature are correct and simply ask how they can be verified, or we take a more ambitious stance and ask if they arise from some deeper level that we have not yet comprehended. The latter is the approach of this chapter. We shall forget absolute space and time and take Platonia for real. I have likened it to a country; countries are there to be explored. In exploring a country, one follows a path through it. Any continuous curve through Platonia is such a path. A natural question is whether some paths are distinguished compared with others. It leads directly to the idea of optimization.

Optimization problems arise naturally, and were already well known to mathematicians in antiquity. It seems they were also known and understood by Queen Dido, who when she came to North Africa was granted as much land as she could enclose within the hide of a cow. She cut it into thin strips, out of which she made a long string. Her task was then to enclose the maximum area of land within it. The solution to this problem of maximizing the area within a figure of given perimeter is a circle.

However, Dido's territory was to adjoin the coast, which did not count as part of the perimeter. For a straight coastline the solution to this problem is a semicircle, and this was said to be the origin of the territory of Carthage. A rich body of mathematical and physical theory has developed out of similar problems. It cannot explain why the universe is, but given that the universe does exist it goes a long way to explain why it is as it is and not otherwise.

In early modern times, Pierre de Fermat (of the famous last theorem) developed a particularly fruitful idea due to Hero of Alexandria, who had sought the path of a light ray that passes from one point to another and is reflected by a flat surface on its way. Hero solved this problem by assuming that light travels at a constant speed and chooses the path that *minimizes* the travel time. Fermat extended this least-time idea to refraction, when light passes from one medium to another, in which it may not travel at the same speed as in the first medium. When a ray of light passes from air into water, the ray is refracted (bent) downward, towards the normal (perpendicular) to the surface. If this behaviour is to be explained by the least-time idea, light must travel slower in water than in air. For a long time it was not known if this were so, so Fermat's proposal was a prediction, which was eventually confirmed.

In 1696 John Bernoulli posed the famous 'brachistochrone' (shortest-time) problem. A bead, starting from rest, slides without friction under gravity on a curve joining two points at different heights. The bead's speed at any instant is determined by how far it has descended. What is the form of the curve for which the time of descent between the two points is shortest? Newton solved the problem overnight, and submitted his solution anonymously, but Bernoulli, recognizing the masterly solution, commented that Newton was revealed 'as is the lion by the claw print'. The solution is the cycloid, the curve traced by a point on the rim of a rolling wheel.

Soon there developed the idea that the laws of motion – and thus the behaviour of the entire universe – could be explained by an optimization principle. Leibniz, in particular, was impressed by Fermat's principle and was always looking for a reason why one thing should happen rather than another. This was an application of his principle of sufficient reason: there must be a cause for every effect. Leibniz famously asked why, among all possible worlds, just one should be realized. He suggested, rather loosely, that God – the supremely rational being – could have no alternative but to create the best among all possible worlds. For this he was satirized as Dr Pangloss in Voltaire's *Candide*. In fact, in his main philosophical work, the *Monadology*, Leibniz makes the more defensible claim that the actual world is distinguished from other possible worlds by possessing 'as much variety as possible, but with the greatest order possible'. This, he says, would be the way to obtain 'as much perfection as possible'.

Inspired by such ideas, the French mathematician and astronomer Pierre Maupertuis (another victim of Voltaire's satire), advanced the *principle of least action* (1744). From shaky initial foundations (Maupertuis wanted to couple his idea with a proof for the existence of God) this principle grew in the hands of the mathematicians Leonhard Euler and Joseph-Louis de Lagrange into one of the truly great principles of physics. As formulated by Maupertuis, it expressed the idea that God achieved his aims with the greatest economy possible – that is, with supreme skill. In passing from one state at one time to another state at another time, any mechanical system should minimize its *action*, a certain quantity formed from the masses, speeds and separations of the bodies in the system. The quantities obtained at each instant were to be summed up for the course actually taken by the system between the two specified states. Maupertuis claimed that the resulting total action would be found to be the minimum possible compared with all other conceivable ways in which the system could pass between the two given states. The analogy with Fermat's principle is obvious.

Unfortunately for Maupertuis's theological aspirations, it was soon shown that in some cases the action would not have the smallest but greatest possible value. The claims for divine economy were made to look foolish. However, the principle prospered and was cast into its modern form by the Irish mathematician and physicist William Rowan Hamilton a little under a hundred years after Maupertuis's original proposal. A wonderfully general technique for handling all manner of mechanical problems on the basis of such a principle had already been developed by Euler and, above all, Lagrange, whose *Mécanique analytique* of 1788 became a great landmark in dynamics.

The essence of the principle of least action is illustrated by 'shortest' paths on a smoothly curved surface. In any small region, such a surface is effectively flat and the shortest connection between any two neighbouring points is a straight line. However, over extended areas there are no straight lines, only 'straightest lines', or *geodesics*, as they are called. As the idea of shortest paths is easy to grasp, let us consider how they can be found. Think of a smooth but hilly landscape and choose two points on it. Then imagine joining them by a smooth curve drawn on the surface. You can find its length by driving pegs into the ground with short intervals between them, measuring the length of each interval and adding up all the lengths. If the curve winds sharply, the intervals between the pegs must be short in order to get an accurate length; and as the intervals are made shorter and shorter, the measurement becomes more and more accurate. The key to finding the shortest path is exploration. Having found the length for one curve joining the chosen points, you choose another and find its length. In principle, you could examine systematically all paths that could link the two chosen points, and thus find the shortest.

This is indeed exploration, and it contains the seed of rational explanation. There is something appealing about Leibniz's idea of God surveying all possibilities and choosing the best. However, we must be careful not to read too much into this. There does seem to be a sense in which Nature at least surveys all possibilities, but what is selected is subtler than shortest and more definite than 'best', which is difficult to define. Nothing much would be gained by going into the mathematical details, and it will be sufficient if you get the idea that Nature explores all possibilities and selects something like a shortest path. However, I do need to emphasize that Newton's invisible framework plays a vital role in the definition of action.

Picture three particles in absolute space. At one instant they are at points A, B, C (initial configuration), and at some other time they are at points A^*, B^*, C^* (final configuration). There are many different ways in which the particles can pass between these configurations: they can go along different routes, and at different speeds. The action is a quantity calculated at each instant from the velocities and positions that the particles have in that instant. Because the positions determine the potential energy, while the velocities determine the kinetic energy, the action is related to both. In fact, it is the difference between them. It is this quantity that plays a role like distance. We compare its values added up along all different ways in which the system could get from its initial configuration to its final configuration, which are the analogues of the initial and final points in the landscape I asked you to imagine. The history that is actually realized is one for which the action calculated in this way is a minimum. As you see, absolute space and time play an essential role in the principle of least action. It is the origin of the two-and-a-bit puzzle. Now let us see how it might be overcome.

DEVELOPING MACHIAN IDEAS

After it became clear to me that Platonia was the arena in which to formulate Mach's ideas, I soon realized that it was necessary to find some analogue of action that could be defined using structure already present in Platonia. With such an action it would be possible to identify some paths in Platonia as being special and different from other paths. In Leibniz's language, such paths could be actual histories of the universe, as opposed to merely possible ones. The problem with Hamilton's action was that it included additional structure that was present if absolute space and time exist, but absent if you insist on doing everything in Platonia. In 1971, with a growing family and financial commitments, I was doing so much translating work I had little time for physics. As luck would have it, the postal workers in Britain went on an extended strike. No more work reached me (no one thought of using couriers in those days) – it was bliss. I got down to the physics and soon had a first idea. It still took quite a time to develop it adequately, but eventually I wrote it up in a paper published in *Nature* in 1974. Mach's principle may be controversial, but it always attracts interest, and *Nature* also published quite a long editorial comment on the paper. Perhaps it was worth waiting ten years before getting my first paper published.

It was certainly a turning point in my life. Some months after it appeared, I received a letter with some comments on it from Bruno Bertotti, who was, and still is, a professor of physics at the University of Pavia in Italy. Bruno, who is a very competent mathematician, has worked in several fields in theoretical physics. In fact, he was one of the last students of the famous Erwin Schrödinger, the creator of wave mechanics (Box 1). But he has also been active in experimental gravitational physics, and he organized the first two – and very successful – international conferences in the field. Although I can never stop thinking about basic issues in physics, I am at best an indifferent mathematician, so I was very lucky that my correspondence with Bruno soon developed into active collaboration. Sometimes Bruno came to work at College Farm, but mostly I went to Pavia. For seven years I went there for about a month, every spring and autumn. It was a very fruitful and rewarding collaboration: my work on Mach's principle would never have developed into a real theory without Bruno's input. I cannot say that we discovered any really new physics, because in the end we had to recognize that Einstein had got there long before us. What I think was important was that in two papers, published in 1977 and 1982, we laid the foundations of a genuine Machian theory of the universe. To our surprise, we then found that this theory is already present within general relativity, though so well hidden that no one (not even Einstein) suspected it. We had found a quite new route to his theory, and had the consolation to know that Einstein had by no means fully grasped the significance of his own theory.

In this connection, a remarkable coincidence that happened to me on my first visit to Pavia is worth recounting. I arrived on a Friday night. I was going to spend the first weekend sightseeing, and after breakfast on Saturday morning I wandered off with no set aim through the streets of Pavia in the warm April sunshine. After about twenty minutes I chanced upon a grand medieval house. A plaque outside said that in the 1820s the poet Ugo Foscolo had lived there. One could walk into the courtyard, which I did. It was Italy as you dream of it. This, I thought, was the place to live. Six months later, quite by chance, I learned that for two years, in the 1890s, it had been Einstein's home. In his teens, the electrical firm run by his father and uncle in Munich had failed, and they had moved to Pavia and started another firm (which also failed). Somehow that chance episode in Pavia seems symbolic of my efforts in physics. Einstein was there first, long ago, but it was still worth the journey to see the place from the inside. It yielded a perspective, quite different from Einstein's, which persuades me that Platonia is the true arena of the universe. If it is, we shall have to think about time differently.

The first idea Bruno and I developed had several interesting and promising properties. Above all, it showed that a mechanics of the complete universe containing only relative quantities and no extra Newtonian framework could be constructed. Hitherto, most people had thought this to be impossible. Just as Mach had suspected, the phenomenon that Newton called inertial motion in absolute space could be shown to arise from motion relative to all the masses in the universe.

We also showed that an external time is redundant. However, besides the desirable features we obtained effects which showed that the theory could not be right. While the universe as a whole could create the experimentally observed inertial effects that we wanted, the Galaxy would create additional effects, not observed by astronomers, that ruled out our approach.

The idea that Bruno and I first developed seemed so natural it surprised us that no one had thought of it earlier. However, I learned quite recently that something similar was proposed in 1904 (in an obscure booklet by one Wenzel Hofmann), and then rediscovered in 1914 by the physicist Hans Reissner and again in 1925 by none other than Schrödinger, just before he discovered wave mechanics. This was especially ironic since Bruno had been Schrödinger's student. I think the main reason why these papers got overlooked was that they were completely overshadowed by Einstein's general relativity and the excitement of the discovery of quantum mechanics in 1925/6. There is also an undoubted tendency for physicists to work within a so-called paradigm (the American philosopher of science Thomas Kuhn's famous expression), and pay at best fleeting attention to ideas that do not fit within the existing established patterns of thought.

I mention these things because the next idea that Bruno and I tried seems to me just as natural as our first idea, if one approaches the problems of describing motion and change with an open mind. It does, however, seem very different from the present paradigm, which has become deep-rooted with the long hegemony of Newtonian ideas, which were only partly changed by Einstein. Although, as we shall see, our second idea is actually built into Einstein's theory at its very heart, within the context of classical physics it merely provides a different perspective on that theory. However, for the study of quantum effects it does represent a genuine alternative, and the attempt to create a quantum theory of the universe may force its adoption, alien though it may appear to many working scientists.

EXPLORING PLATONIA

Let me now explain this second idea. So far, I have explained only what the points of Platonia are. Each is a possible relative arrangement, a configuration, of all the matter in the universe. If there are only three bodies in it, Platonia is Triangle Land, each point of which is a triangle (Figures 3 and 4). Can we somehow say 'how far apart' any two similar but distinct triangles are? If so, this will define a 'distance' between neighbouring points in Triangle Land, and just as mathematicians seek geodesics using ordinary distances on curved surfaces, we can start to look for geodesics in Platonia. If we can find them, they will be natural candidates for actual histories of the universe, which we have identified as paths in Platonia. They will be Machian histories if the 'distance' between any two neighbouring points in Platonia is determined by their structures and nothing else, and we shall not need to suppose that they are embedded in some extra structure like absolute space.

There is such a simple and natural solution to the problem of finding geodesics in Platonia that I would like to spell it out. The fact that it does seem to be used by nature is one of the two prime pieces of evidence I have for suggesting that the universe is timeless. (The second, equally simple in its way, comes from quantum mechanics.) How it works out for the simplest example of a universe of three bodies is described in Box 8.

BOX 8   Intrinsic Difference and Best Matching

In Figure 21, triangle ABC is one point in Triangle Land, and the slightly different triangle $A^*B^*C^*$ is a neighbouring point. A 'distance' between them can be found in many ways, but one of the simplest is the following. Imagine that ABC is held fixed, and $A^*B^*C^*$ is placed in any position relative to it. This creates 'distances' AA^*, BB^*, CC^* between the corresponding vertices, at which we suppose there are bodies of masses a, b, c. We form a 'trial distance' d by taking each mass and multiplying it by the square of the corresponding distance, adding the results and taking the square root of the sum. Thus

$$d = \sqrt{a\,(AA^*)^2 + b(BB^*)^2 + c(CC^*)^2}$$

This is an arbitrary quantity, since the relative positioning of the two triangles is arbitrary. It is, however, possible to consider all relative positionings and find the one for which d is minimized. This is a very natural quantity to find, and it is not arbitrary. Two different people setting out to find it for the same two triangles would always get the same result. It measures the *intrinsic difference* between the two matter distributions represented by the triangles. It is completely determined by them, and does not rely on any external structure like absolute space.

The intrinsic difference between two arbitrary matter distributions can be found similarly. One distribution is supposed fixed, and the other moved relative to it. In any trial position, the analogue of the above expression is calculated, and the position in which it is minimized is sought. Because this special position reduces the apparent difference between the two matter configurations to a minimum, it may be called the *best-matching position*.

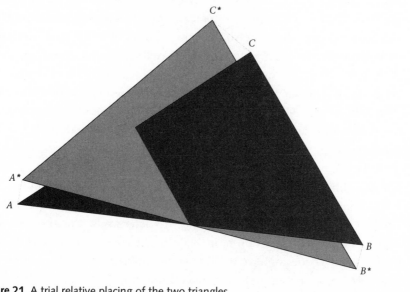

Figure 21 A trial relative placing of the two triangles.

Using the intrinsic difference defined in Box 8, we can determine 'shortest paths' or 'histories' in Platonia as explained above. However, the intrinsic difference by itself does not lead to very interesting histories, and it is more illuminating to consider a related quantity. The potential energy of any matter distribution (Figure 17) is determined by its relative configuration, and is therefore already 'Machian'. Each matter distribution has its own Newtonian gravitational potential energy. Two nearly identical matter distributions have almost the same potential. Now, the intrinsic difference is determined by two nearly identical configurations. To obtain more interesting histories we can simply multiply the intrinsic differences by the potential (strictly speaking, by the square root of minus the potential). This will change the definition of 'distance', but it will still enable us to determine 'shortest distances'. You do not need to worry about these details, but I do want to give you a flavour of what is involved.

I think you will agree that finding shortest paths in an imagined time-less landscape bears little direct resemblance to our powerful sense of the passage of time. Yet the outcome turned out to be remarkably like what happens in Newtonian theory. Let me explain, taking again the example of a three-body universe, for which Platonia is Triangle Land.

Any continuous path in it corresponds to a sequence of triangles: they are the 'points' through which the path passes. But this is very similar to what comes out of Newtonian theory (Figure 1) – which, however, yields not only the triangles but also their positions in absolute space and separations in time. Remember that the triangles in Figure 1 are 'lit up' by flashes at each unit of absolute time, and that we see them, in perspective, in absolute space at those times. However, these are invisible aids. The astronomers see neither when they look through telescopes, all they see are stars. Thus, as far as observable things are concerned, both theories yield the same kind of thing – sequences of triangles. The question is, what kind of sequences do the two theories yield? In what respects do the theories differ when it comes down to what is actually observable?

The major difference is that the Machian theory makes more definite predictions. As a theory of geodesics, it determines the shortest path between any two fixed points in Platonia. It covers Platonia with such paths. These geodesics have the following important property: at any one point in Platonia, many of them pass through it. In fact, for every direction that you can go from a point, there is just one geodesic. This is the crucial difference. As Figures 13 and 14 highlighted, when the Newtonian histories are represented as paths in Platonia, it turns out that many can pass through the same point, and have at that point the same direction. However, these paths then 'splay out' and go to quite different places in Platonia. In Newtonian terms, they differ in energy and angular momentum. The difference is not apparent in the initial point and direction, but comes to light dramatically in the later evolution of the paths. This defect is absent in the Machian theory. For any given point and direction at that point, there is just one geodesic. Bruno and I constructed the theory precisely to achieve that aim.

What did quite surprise us was to discover that the unique Machian history with a given direction through a point is identical to one of the many Newtonian histories through the point with the same direction. It is, in fact, the Newtonian history for which the energy and angular momentum are both exactly zero. The small fraction of Newtonian solutions with this property are all the solutions of a simpler timeless and frameless theory.

This brought to light an unexpected reconciliation between the positions of Newton and Leibniz in their debate about absolute and relative motion. Both were right! The point is that in a universe which, like ours, contains many bodies, there can be innumerable subsystems that are effectively isolated from one another. This is true of the solar system within the Galaxy, and also for many of the galaxies scattered through the universe. Each subsystem, considered by itself, can have non-zero energy and angular momentum. However, if the universe is finite, the individual energies and angular momenta of its subsystems can add up to zero. In a universe governed by Newton's laws this would be an implausible fluke. But if the universe is governed by the Machian law, it must be the case. It is a direct consequence of the law. What is more, the Machian law predicts that in a large universe all sufficiently isolated systems will behave exactly as Newton predicted. In particular, they can have non-zero energy and angular momentum, and therefore seem to be obeying Newton's laws in absolute space and time. But what Newton took to be an unalterable absolute framework is shown in the Machian theory to be simply the effect of the universe as a whole and the one law that governs it. What physicists have long regarded as laws of nature and the framework of space and time in which they hold are, as I said in Chapter 1, both 'local imprints' of that one law of the universe.

You can see directly how absolute space and time are created out of timelessness. Take some point on one of the Machian geodesics in Platonia; it is a configuration of masses. Take another point a little way along the geodesic; it is a slightly different configuration. Without any use of absolute space and time, using just the two configurations, you can bring the second into the position of best matching relative to the first. You can then take a third configuration, a bit farther along the path, and bring it into its best matching position relative to the second configuration. You can go along the whole path in this way. The entire string of configurations is oriented in a definite position relative to the first configuration. What looks like a framework is created, but it is not a pre-existing framework into which the configurations of the universe are slotted: it is brought into being by matching the configurations. Nevertheless, we get something like the Newtonian picture in Figure 1, except that we do not as yet have the 'spacings in time'.

But this too emerges from the Machian theory. In the equations that describe how the objects move in the framework built up by best matching, it is very convenient to measure how far each body moves by making a comparison with a certain average of all the bodies in the universe. The choice of the average is obvious, and simplifies the equations dramatically. No other choice does the trick. For this reason it needs a special name; I shall call it the *Machian distinguished simplifier*. It is directly related to the quantities used to determine the geodesic paths in Platonia. To find how much it changes as the universe passes from one configuration to another slightly different one, it is necessary only to divide their intrinsic difference by the square root of minus the potential. (The action, by contrast, is found by multiplying it by the same quantity.) When this distinguished simplifier is used as 'time', it turns out that each object in the universe moves in the Machian framework described above exactly as Newton's laws prescribe. Newton's laws and his framework both arise from a single law of the universe that does not presuppose them.

In such a universe, the ultimate standard of time that determines which curve is traced by Galileo's ball when it falls off his table in Padua is unambiguous. It is the average of all the changes in the universe that defines the Machian distinguished simplifier. Time is change, nothing more, nothing less.

The difference between the Newtonian and Machian theories can be summarized as follows. If we do not know the energy and angular momentum of a Newtonian system, we always need at least three snapshots of its configurations in order to reconstruct the framework of space and time in which they obey Newton's laws. The task is complicated, to say the least. If, however, the system is Machian, the framework can be found with just two snapshots and the task is vastly simpler. It simply requires best matching of the two configurations.

When, later, I suggest that the quantum universe is timeless in a deeper sense than the classical Machian universe just described, that will be a conjecture. But it is made plausible by the results of this chapter. They are not speculation but mathematical truths. Every phenomenon explained by Newton's laws, including the beautiful rings of Saturn and the spectacular structure of spiral galaxies, can be explained without absolute space and time. They follow from a simpler, timeless theory in Platonia.

PART 3

The Deep Structure of General Relativity

Now we come to relativity. My aim is not to give an extended account, only to show how its fundamental features relate to the book's theme. But I have a tough nut to crack. My subject is the non-existence of time, whereas time is almost everything in relativity as it is usually presented. Is relativity *Hamlet* without the Prince of Denmark?

In fact, the evidence for the non-existence of time in relativity has long been hidden by accidents of historical development, and is far stronger than many people realize. Yet the case is not quite conclusive. We have seen how the space and time of Newton's theory can be constructed from instants of time as defined in this book. Taking them to be the true atoms of existence, we have shown that no external framework is needed. Einstein's space-time can also be put together from instants in a strikingly similar way. However, in the finished product they are knit together far more tightly than in Newtonian theory. Explaining the wonderful way in which this happens is the goal now. If the world were classical, no one would try to pull space-time apart into instants. But quantum theory will probably shatter space-time. It is therefore sensible to consider the constituents into which it might shatter. This is what I shall do in Part 3.

I begin by looking at the special theory of relativity, in which gravity plays no role. I then go on to the general theory, in which Einstein found a most brilliantly original way to describe gravity. In both relativity theories time seems to be very real and to behave in baffling ways. But, as became clear only after Einstein's death, his theory has a deep structure which is revealed only by an analysis of how it works as a dynamical theory. It is this deep structure that is timeless. Quite a large proportion of Part 3 will explain the purely historical accidents that obscured the deep structure of general relativity for so long.

CHAPTER 8

The Bolt from the Blue

HISTORICAL ACCIDENTS

In the whole of physics, nothing is more remarkable than the transformation wrought by a simple question that Einstein posed in 1905: what is the basis for saying that two events are simultaneous? Einstein was not the first to ask it. James Thomson, brother of Lord Kelvin, had in 1883. More significantly, so had Poincaré – a great figure in science – in 1898, in a paper that Abraham Pais, Einstein's biographer, calls 'utterly remarkable'. In connection with historical accidents, Poincaré's paper is very interesting. He identified *two* problems in the definition of time.

First he considered duration: what does it mean to say that a second today is the same as a second tomorrow? He noted that this question had recently been widely discussed, and he outlined the astronomers' solution, the ephemeris time described in Chapter 6. However, he then noted a second question, just as fundamental and in some ways more immediate, which had escaped close attention. How does one define simultaneity at spatially separated points? This was the question that Einstein posed and answered seven years later with such devastating effect. I read the subsequent history of relativity as follows. Einstein answered his question – Poincaré's second – with such aplomb and originality that it eclipsed interest in the question of duration. It is not that duration plays no role in relativity – quite the opposite, it plays a central role. But duration is not derived from first principles. It appears indirectly.

One of the main aims of Part 3 is to redress the balance, to treat duration at the same level as simultaneity. There is, in fact, a beautiful theory of duration at the heart of general relativity, but it is hidden away in sophisticated mathematics. Einstein had no inkling of this. He said of his own theory that no one who had grasped its content could 'escape its magic'. But the magic was more potent than even he realized. It can, it almost certainly will, destroy time.

BACKGROUND TO THE CRISIS

Much of nineteenth-century physics can be seen as meticulous preparation for the denouement over simultaneity. It had to come, but what a *coup de théâtre* Einstein made of it. Many readers will be familiar with the story, but since it introduces important ideas I shall briefly recall some key elements. It all started with an investigation of interference carried out in 1802 by the English polymath Thomas Young, famous among other things for his decipherment of the Egyptian hieroglyphs on the Rosetta Stone. In a sense, this was the start of both relativity and quantum theory. Young observed that if light from a single source is split into two beams that are subsequently recombined and projected onto a screen, then bright and dark fringes appear. He interpreted them in terms of a wave theory of light. If light is some kind of wave motion, there will be wave crests and troughs in both beams. When they are recombined, there will be places where the crests from one beam coincide with troughs in the other. They will cancel, giving dark fringes. But where crests coincide, they will enhance each other, giving bright fringes (Figure 22). Innumerable natural phenomena are explained by interference.

Young's insight, which was developed more or less independently and much more thoroughly some years later by the Frenchman Augustin Jean Fresnel, soon gave rise to the notion that light waves must be vibrations of some elastic medium, which was called the *aether*. Meanwhile, the study of electricity and magnetism developed rapidly. In 1831, the English scientist Michael Faraday discovered electromagnetic induction, which not only showed that electricity and magnetism were related phenomena but rapidly became the basis of all electrical machinery. Deeply impressed by the patterns formed by iron filings sprinkled on paper held near a magnet (Figure 23), Faraday introduced the notion of lines of force and *fields*. A field can be thought of as a tension or excitation that exists throughout space and varies continuously (as demonstrated by induction) in both space and time. The field concept eventually changed physicists' picture of what the world is 'made of'.

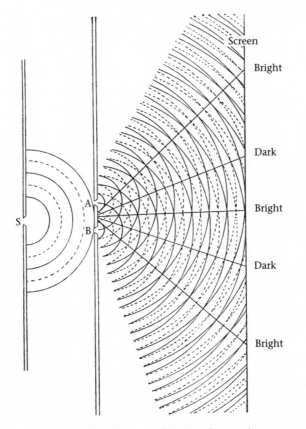

Bright

Dark

Bright

Dark

Bright

Screen

Figure 22 Thomas Young's original explanation of the interference fringes in accordance with the wave theory of light, which he deduced by analogy with the behaviour of water waves. According to this interpretation, the beam reaches the barrier in the form of a plane wave, the successive parallel crests of which arrive simultaneously at the two slits A and B. The wave is diffracted at each slit, and spherical waves spread out from each point of the two slits towards the screen. At some points on the screen, the wave crests (or troughs) from the two slits arrive simultaneously, and the wave intensity is enhanced (bright regions). At other points, a wave crest from one slit arrives with a wave trough from the other. The wave intensity is cancelled at such a point (dark regions). This is the classical explanation of the fringes in terms of interference.

In the decade from 1855, the Scottish physicist James Clerk Maxwell took up Faraday's qualitative field notion and cast it into mathematical form. His equations showed that electromagnetic effects should propagate through empty space as waves with a speed determined by the ratio of certain constants. It had already been noted that the ratio was equal to the known speed of light, leading to the strong suspicion that light was an electromagnetic effect. Maxwell's equations proved this. Electromagnetic

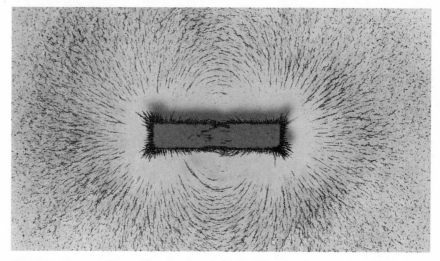

Figure 23 Magnetic lines of force as revealed by placing iron filings in the magnetic field of a bar magnet.

effects can propagate as waves of many different wavelengths: from radio waves (with wavelengths of around a metre to a kilometre), microwaves (wavelengths measured in centimetres), infrared waves (some ten to a thousand waves per centimetre), visible light (roughly ten thousand waves to the centimetre), ultraviolet light (up to around a million waves per centimetre), X-rays (of the order of ten million waves in every centimetre) and gamma rays (billions or even trillions of waves per centimetre). Hertz's celebrated detection of waves from an electromagnetic source in 1888 was the first confirmation of this consequence of Maxwell's theory.

Virtually all physicists were convinced that these electromagnetic excitations must be carried by some mechanical aether. This put a remarkable twist into the theory of motion and the status of Newton's absolute space. Even in the framework of Newtonian theory, there had always been one serious problem with the notion. Newton was not entirely frank about it. In his guts, he certainly believed in a state of absolute rest. When he introduced absolute space, his words suggested the existence of one unique framework of motion. Either you moved with respect to it or you did not. However, in the body of the *Principia* he stated and used correctly the relativity principle, according to which the motions within a system are completely unaffected by any uniform overall motion it has (Box 5). This seriously undermined the idea of a unique state of rest – no criterion could establish whether one were in it or moving uniformly.

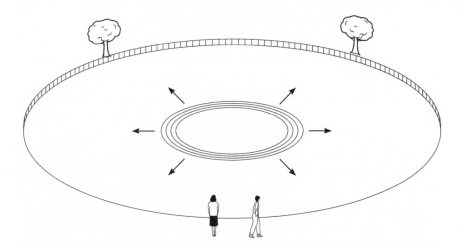

Figure 24 The pond argument. According to the spectator standing on the bank, the ripples move to the left and right with equal speed. But her partner, walking along the bank, sees things differently. He can almost keep up with the waves going to the right, while the left-moving waves recede from him almost twice as fast.

It was soon seen that the aether should introduce an experimentally verifiable standard of rest. The argument is simple and seems irrefutable (Figure 24). If you throw a stone onto the still surface of a pond, waves spread out in concentric rings. The water molecules do not move with the wave, they merely go up and down. The water plays the role of the conjectured aether: itself at rest, it is the material carrier of the waves. As seen by the woman standing on the bank, the waves spread out in all directions with the same speed. But for her partner walking along the bank, the wave process unfolds differently. The waves moving in the same direction as him have a different velocity relative to him compared with the waves moving in the opposite direction. A fast walker will even overtake some of the waves. The relativity principle cannot hold for such processes, and it was therefore expected that it would hold only for the mechanical processes described by Newton's laws, but not for optical and electromagnetic phenomena. Moreover, as the Earth revolves in its orbit around the Sun it must be continually changing its speed through the aether. This ought to result in observable effects.

In fact, the argument is not quite so simple. Everyone agreed that light must be carried by an aether, but were all parts of the aether at rest relative to one another? As the Earth orbits the Sun, might it not carry some aether with it? It was also necessary to consider what would happen to light passing through water flowing on the Earth. Would the aether be carried along, partially or completely, by the water relative to the Earth? In fact, many issues had to be considered, including aberration, which makes the stars seem to shift slightly towards the point in the sky towards which the Earth is moving at any instant as it orbits the Sun. The arguments, some of which predated Maxwell's work, developed over a period of eighty years, and many important experiments were made. By 1895, when the Dutch physicist Hendrik Anton Lorentz published an influential study, a consensus had more or less developed. It was that all known experimental results, with one crucial exception, could be explained by assuming the existence of a perfectly rigid aether.

The aether as proposed by Lorentz was actually devoid of all physical properties except rigidity. It was simply there to carry the excitations of Maxwell's electromagnetic field and, in Lorentz's words, to be the framework 'relative to which all observable motions of the celestial bodies take place'. It therefore supplied a standard of rest like the water in the pond.

The one exception with which Lorentz had to contend was the famous Michelson–Morley experiment performed with great accuracy in 1887. Based on interference between light beams moving in the direction of the Earth's motion and at right angles to it, it was designed to measure the change in the speed of the Earth's motion through the aether over the course of a year. Its accuracy was sufficient to detect even only one-hundredth of the expected effect. But nothing was observed. It was a great surprise, and very puzzling.

Lorentz's response was piecemeal. In particular, he suggested that, for some physical reason, the length of a body moving relative to the aether could be reduced in the direction of its motion by the amount needed to explain the Michelson–Morley result. Some years earlier, the Irish physicist George Fitzgerald had made the same proposal. Poincaré responded that some general principle should rule out all possibility of detecting motion relative to the aether. It should not be necessary to invoke ad hoc hypotheses. He began to think that the relativity principle might hold universally and not just for mechanical phenomena. Both he and Lorentz were working in this direction when Einstein appeared on the scene with a stunning solution.

EINSTEIN AND SIMULTANEITY

Two aspects of Einstein's work ensured its triumphant success. First, he took the relativity principle utterly seriously. It was the bedrock, repeatedly exploited. Second, he took for real a 'local time' that Lorentz had introduced as a formal device to describe phenomena in a reference frame moving relative to the aether. Events simultaneous in the 'local time' were not so in the real time of the aether frame. But Einstein, committed to relativity, regarded one as just as real as the other. He made a virtue out of an apparent vice, and saw that the key to the entire mystery lay in the concept of simultaneity.

He deliberately highlighted an apparently irreconcilable paradox and then deftly presented its unique resolution: a radical proposal for saying when events are simultaneous. Hitherto this had seemed obvious, but Einstein showed that simultaneity was not a property of the world but a reflection of the way we describe it. By showing that the paradox could be resolved only by changing the notion of simultaneity – and with it time – he brought this issue to the fore.

The paradox was carefully prepared. He first defined the relativity principle. As in mechanics (Box 5), he postulated distinguished frames of reference in which all the laws of nature take their simplest form, and required this form to be the same in each frame. Any such frame, which constituted a kind of grid in space and time, should be in a state of uniform rectilinear (i.e. straight) motion relative to any other. He then postulated, in addition to this general principle, just one actual law of nature: that light propagates with the same speed c in all directions irrespective of the speed of the source. This was exactly what everybody had always assumed would hold in the unique frame of reference at rest in the aether. Einstein insisted it should happen in all frames.

The pond argument suggests that this is absurd. But Einstein realized that he had a hitherto unrecognized freedom: the grid lines defining simultaneity in space and time could be 'drawn' in a novel way. Simultaneity at spatially separated points must be defined in some way – but how? There must be a physical transmission of signals so as to synchronize clocks. The ideal would be infinitely fast signals. Then there would be no argument. This is effectively what happens in the pond experiment – the man and woman observe the water waves by light, which travels nearly a billion times faster than they do. We now see that the analogy between water waves in ponds and light waves in the aether is not perfect. For a full analogy, there would have to be signals that travel faster than light itself.

But such signals were unknown in Einstein's time, and his theory would show that they could not exist. He therefore used the best substitute – light. This completely changed things. Light was to be analysed in a framework that light itself created, so the problem became self-referential. It might seem that Einstein cheated, making up the rules as he went along to ensure that he won the game. However, he was simply confronting a fact of life: laws of nature will be meaningful only if they relate things that can actually be observed. We live inside, not outside, the universe, and to synchronize distant clocks we have no alternative to the physical means available to us. Einstein's hunch that we should use light because it would turn out to be the fastest medium available in nature has so far been totally vindicated.

The magical touch was that his choice was the most natural thing to do – in the theory of an aether *and* in the context of the relativity principle. Given their apparent irreconcilability, his subsequent demonstration of their compatibility was a coup. It also showed that there was something inevitable about the result.

For suppose there is a pond-like aether and that nothing is faster than light. It is natural to assume that it travels equally fast in all directions. Then how are we to define simultaneity throughout the aether? Einstein proposed setting up a master clock at a central reference point, sending a light signal to some distant identical clock at rest relative to it, and letting the signal be reflected back to the master clock. If it measures a time T for the round trip, we would obviously say that the light took $\frac{1}{2}T$ to reach the distant clock, which can be synchronized to read $t + \frac{1}{2}T$ on the arrival of a signal sent by the master clock when it reads t. In this way, clocks throughout the aether can be synchronized with the master clock. Standard measuring rods can be used to measure the distances between them. This is the obvious way to set up a space-time grid if the aether theory is correct.

However, it does assume that the aether is 'visible' and that we know when we are at rest in it. But this the relativity principle denies. Imagine a family of observers, equipped with clocks to measure time and rods to measure length, distributed in space and at rest relative to one another.

Believing themselves at rest in the aether, they define simultaneity by Einstein's prescription. There is also a second family, with identical rods and clocks, also at rest relative to one another but moving uniformly relative to the first. By the relativity principle, they can equally believe themselves at rest in the aether. So they too will use Einstein's prescription to define simultaneity. Just as belief in the aether theory makes the prescription natural, belief in the relativity principle makes it natural for both families to adopt it. Nothing in nature privileges one family over the other. Whatever one family does, the other can do with equal right. In particular, each can use Einstein's prescription.

The inescapable consequence is that the two families will disagree about which events are simultaneous. However, by accepting this, Einstein achieved his first goal – the demonstration that light propagation takes an identical form for both families (Box 9). This remarkable fact – that the relativity principle holds for light propagation and that simultaneity depends on the observer and on convention – is thus the great denouement towards which so much wonderful physics in the nineteenth century had been tending. It also showed that the aether is a redundant concept, since no experiment can establish whether we are moving relative to it.

Lack of simultaneity was only the beginning. Einstein went on to draw further amazing consequences from his iron insistence that all phenomena must unfold in exactly the same way for any two families of observers in uniform motion relative to each other. In particular, he was able to make some startling predictions about rods and clocks. The point is that the facts of light propagation are established by means of physical rods and clocks, but these tools are not immune to the relativity principle. Using simple equations and precise arguments, Einstein showed that two such families must each come to the conclusion that the clocks of the other family, moving relative to them, run slower than their own clocks. Each family also concludes that the rods of the other family are shorter than their own.

What is so remarkable about these results – and it seems so impossible that many quite intelligent people still refuse to accept it – is their mutual nature. How can it be that each family finds that the clocks of the other family run slower than their own? Box 10 explains.

BOX 9   Relativity in One Diagram and 211 Words

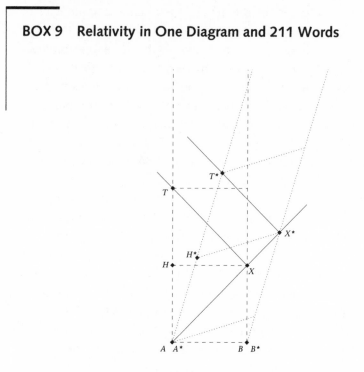

Figure 25 Alice (A) and Bob (B) believe they are at rest in the aether, and therefore draw a grid (dashes) with time vertical and space (only one dimension shown) horizontal. To synchronize their clocks, Alice sends Bob a light signal (solid line), which reaches him at X, where it is reflected back to Alice at T. Alice concludes that the signal reached Bob when she was at H. Their identical twins Alice* (A*) and Bob* (B*) are moving uniformly relative to them, but also believe they are at rest. Alice* sends a light signal, just as her twin does, at the moment they meet. It reaches Bob* at X*, and the reflection of it returns to Alice* at T*. She therefore concludes that her signal reached Bob* at H*, so she and Bob* have a grid (dots) inclined relative to their twins' grid. The pairs do not agree on which events are simultaneous. Alice and Bob think H and X are, their twins think H* and X* are. However, they confirm the relativity principle, since both find that light travels along rays parallel to the diagonals (AX, XT and A*X*, X*T*) of their respective coordinate grids. Despite appearances, the situation is symmetric – in Alice* and Bob*'s grid their twins' grid appears skew.

THE FORGOTTEN ASPECTS OF TIME

Fascinating as the results of Einstein's two relativity theories are, many of them are not directly relevant to my main theme. Popular accounts that cover topics I omit are recommended in the section on Further Reading.

BOX 10   The Impossible Becomes Possible

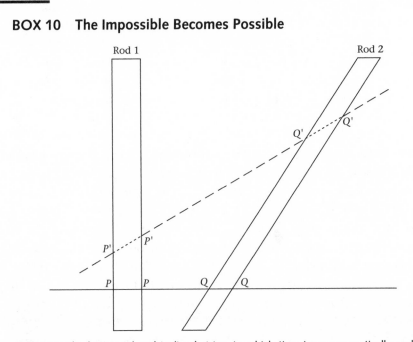

Figure 26 The horizontal and inclined strips, in which time increases vertically and the horizontal represents one space dimension, show the histories of two physically identical rods moving uniformly relative to each other. For Bob and Alice, points on the continuous line *PPQQ* are simultaneous and show the positions and lengths of the rods at the corresponding times. Their rod, *PP*, appears longer than Bob* and Alice*'s rod, *QQ*. But the starred twins think that points on the line *P'P'Q'Q'* are simultaneous, and conclude that their rod *Q'Q'* is longer than *P'P'*. A similar illustration could be given for clocks. Such diagrams do not explain this behaviour of rods and clocks, but do show that there is no outright logical contradiction. Einstein's conclusions are as secure as his premises. His confidence in them has so far been totally vindicated.

My aim in Part 3 is to show how Einstein's approach to relativity led him to an explicit theory of simultaneity but an implicit theory of duration. It is the latter that is important for this book, but it never got properly treated in relativity. The point is that remarkable facts about duration, as revealed through the clocks of different observers, follow inescapably from the definition of simultaneity and the relativity principle. Einstein did not need to create a theory of clocks and duration from first principles in order to learn some facts about them: they already followed from his two primary postulates.

The method Einstein used to create his relativity theories is an important factor. During the nineteenth century, mainly through the development of thermodynamics, physicists began to distinguish between, on the one hand, theories of the world in terms of truly basic laws and constituents (e.g. atoms and fields) and, on the other hand, so-called principle theories. In the latter, no attempt would be made to give an ultimate theory of things. Instead, the idea was to seek principles that seemed to hold with great generality and include them in the foundations of the description of phenomena. The repeated failures of all attempts to build perpetual-motion machines, of which two distinct types could be envisaged, became the basis of the first and second laws of thermodynamics. In the form in which it was developed on this basis, thermodynamics was a theory of the second kind – a principle theory.

In contrast, Lorentz's combined theory of the electromagnetic field, electric charges and the aether was basically a theory of the first kind – it aspired to a fundamental description of the world in terms of its ultimate constituents. Einstein deliberately decided not to follow such a path in his own work on electrodynamics, from which the special theory of relativity emerged. He based it as far as possible on general principles. The fact is that Max Planck's quantum discoveries (Box 1) and Einstein's own development of them a few months before the relativity paper had persuaded Einstein that something very strange was afoot. Despite his admiration for Maxwell's equations, he felt sure that they could not be the true laws of electromagnetism because they completely failed to explain the quantum effects. He had no confidence in his ability to find correct alternatives. Then, and to the end of his days, Einstein found the quantum baffling. He felt deeply that it was a huge mystery. By comparison, relativity (the special theory at least) was almost child's play.

It was this attitude that largely shaped Einstein's strategy in approaching the problem of the electromagnetic aether. He resolved to make no attempt at a detailed description of microscopic phenomena. Instead, he would rely on the relativity principle, for which there seemed to be strong experimental support, and make as few additional assumptions as possible. In the event, he was able to limit these to his assumption about the nature of light propagation. This was the one part of Maxwell's scheme that he felt reasonably confident would survive the quantum revolution.

This had important consequences for the theory of time. Poincaré's 1898 paper showed that it must answer two main questions: how simultaneity is to be defined, and what duration is. Associated with the second question is another, almost as important: what is a clock? Because of his approach, Einstein answered only the first question at a fundamental level. He gave at best only partial answers to the other two, and gave no explicit theories of either rods or clocks. Instead, he tacitly assumed the minimal properties they should possess. Otherwise, he relied to a very great extent on the relativity principle. It took him far. Few things in physics are more beautiful than the way he postulated the universal relativity principle and the one particular law of light propagation, and then deduced, from their combination, extraordinary properties of rods, clocks and time. If the premises were true, rods and clocks had to behave that way.

During his protracted creation of general relativity, Einstein used this trick several times. The strategy was always to avoid specific assumptions, and instead to seek principles. In this way he avoided ever having to address the physical working of rods and clocks: they were always treated separately as independent entities in both relativity theories. Their properties were not deduced from the inner structure of the theory, but were simply required to accord with the relativity principle. Einstein was well aware that this was ultimately unsatisfactory, and said so in a lecture delivered in 1923. He made similar comments again in 1948 in his *Autobiographical Notes*.

However, the tone of his comments does not suggest that he expected any great insight to spring from the rectification of this 'sin' (Einstein's own expression). Only a 'tidying up' operation was needed. This gap in the theory of duration and clocks has still not been filled. I know of no study that addresses the question of what a clock is (and how crucially it depends on the determination of an inertial frame of reference) at the level of insight achieved in non-relativistic physics by James Thomson, Tait and Poincaré. Throughout relativity, both in its original, classical form and in the attempts to create a quantum form of it (which we come to in Part 5), clocks play a vital role, yet nobody really asks what they are. A distinguished relativist told me once that a clock is 'a device that the National Bureau of Standards confirms keeps time to a good accuracy'. I felt that, as the theorist, he should be telling them, not the other way round.

The truth is that a chapter of physics somehow never got written. Despite his great admiration for Mach, Einstein was curiously insensitive to the issues highlighted by Mach and Poincaré. He did not directly address the nature and origin of the framework of dynamics. Despite an extensive search through his published papers and published and unpublished correspondence, I have found no indication that he ever thought really seriously about issues like those raised by Tait's problem. This is rather surprising, since these were 'hot topics' during the very period in which Einstein created special relativity. He did not ask how the spatiotemporal framework (i.e. the framework of space and time used by physicists) arose; instead, he described the finished product and the processes that take place within the arena it creates.

In fact, Einstein and Hermann Minkowski, whose work will shortly be considered, brought about a marked change of emphasis in physics. To use an expression of John Wheeler, the 'royal high road of physics' from Galileo until Einstein was dynamics. Maxwell saw his own work as an extension of the principles developed by Galileo and Newton to new phenomena and to the field notion introduced by Faraday. At the same time, other scientists like Carl Neumann and Mach became aware of the need for new foundations of dynamics. In Poincaré's writings of around 1900, one can see clear hints of how dynamics might have been developed further as the main stream of research. In particular, an explicit theory of the origin of the spatiotemporal framework might have emerged. That is more than evident from Poincaré's 1898 paper on time and his 1902 comments, discussed in Chapter 5.

All this was changed by Einstein's 1905 paper. Because of his quantum doubts, he distrusted explicit dynamical models. Within a few years a dualistic scheme appeared. Newton's absolute space and time were replaced by space-time, but this was not the complete story. Actual physics emerged only with the statements about how rods and clocks behaved in space-time. This is where the scheme was dualistic. The behaviour of rods and clocks – and with it a theory of duration – never emerged organically from the structure of space-time, it was simply postulated. This is not to say the dualistic scheme is wrong in the statements it makes. Einstein's theory is as secure as its foundations; there is no hint of failing there. However, insight into the nature of time and duration was lost.

For all that, general relativity does contain, hidden away in its mathematics (as I have already indicated), a theory of duration and the spatiotemporal framework. However, this did not come to light for many decades and even now is not properly appreciated. How this came about, and an account of the 'hidden dynamical core' of general relativity, are the subject of the next chapters.

It may help to end this chapter with a general remark on time. It is impossible to understand relativity if one thinks that time passes independently of the world. We come to that view only because change is so all-pervasive and so many different changes all seem to march in perfect step. Relativity is not about an abstract concept of time at all: it is about physical devices called clocks. Once we grasp that, many difficulties fall away. If light did not travel so much faster than normal objects, we would observe relativistic effects directly and they would not strike us as strange. There is nothing inherently implausible in the idea that clocks travelling past us at high speed should be observed to go slower than the watch on our wrist. Motion of the clock might well alter the rate at which it ticks. After all, when we swim through water, we feel the way our body responds. If there were an aether, clocks could well be affected by their motion through it. What is difficult to grasp is how observers travelling with the moving clocks think our wristwatch is running slow, while we think just the same about their clocks (this apparent logical impossibility has been dealt with in Box 10). However, the important thing is to get away from the idea that time is *something*. Time does not exist. All that exists are things that change. What we call time is – in classical physics at least – simply a complex of rules that govern the change.

CHAPTER 9

Minkowski the Magician

THE NEW ARENA

Hermann Minkowski's ideas have penetrated deep into the psyche of modern physicists. They find it hard to contemplate any alternative to his grand vision, presented in a famous lecture at Cologne on 21 September 1908. Its opening words, a magical incantation if ever there was one, are etched on their souls:

> The views of space and time which I wish to lay before you have sprung from the soil of experimental physics, and therein lies their strength. They are radical. Henceforth space by itself, and time by itself, are doomed to fade away into mere shadows, and only a kind of union of the two will preserve an independent reality.

The branch of knowledge that considers what exists is ontology. These three sentences changed the ontology of the world – for physicists at least.

For most physicists in the nineteenth century, space was the most fundamental thing. It persisted in time and constituted the deepest level in ontology. Space, in turn, was made up of points. They were the ground of being, conceived as identical, infinitesimal grains of sand close-packed in a block. Space was like glass. It was, of course, three-dimensional. However, alerted by Einstein's work to how the relativity principle mixed up space and time, Minkowski commented that 'Nobody has ever noticed a place except at a time, or a time except at a place.' He had the idea that space and time belonged together in a far deeper sense than anyone had hitherto suspected. He fused them into *space-time* and called the points of this four-dimensional entity *events*. They became the new ground of being.

Such atoms of existence – the constituent events of space-time – are very different from the entities that I suggested in Part 2 as the true atoms of existence. The main aim of Part 3 is to show that space-time can be conceived of in two ways – as a collection of events, but also as an assemblage of extended configurations put together by the principle of best matching and the introduction of a 'time spacing' through a distinguished simplifier, as explained for the Newtonian case at the end of Chapter 7. However, reflecting the relativity of simultaneity, the assemblage has an additional remarkable property that gives rise to the main dilemma we face in trying to establish the true nature of time.

FROM THREE TO FOUR DIMENSIONS

In itself, the fusion of space and time was not such a radical step. It can be done for Newtonian space and time. To picture this, we must suppose that ordinary space has only two dimensions and not three. We can then imagine space as a blank card, and the bodies in space as marks on it. Any relative arrangement of these marks defines an instant of time.

The solution of Tait's problem showed how relative configurations can, if their bodies obey Newton's laws, be placed in absolute space at their positions at corresponding absolute times. If space is pictured as two-dimensional, absolute space is modelled not by a room but by a flat surface. The solution of Tait's problem places each card on the surface in positions determined by the marks on the cards. In these positions, in which the centre of mass can be fixed at one point, any body moving inertially moves along a straight line on the surface.

Keeping all the cards horizontal (parallel to the surface), we can put a vertical spacing between them which is proportional to the amount of absolute time between them. This is like imagining the amount of time between 11 o'clock and 12 o'clock as a distance, and is a very convenient way of visualizing things. The resulting structure can be called *Newtonian space-time*. The one dimension of time has been put together with the two of space. Newton's laws can be expressed very beautifully in this three-dimensional structure, which is a kind of block. Whatever motion a body has, it must follow some path in this block. Minkowski called this path its *world line*. If the body does not move in space, which is a special case of inertial motion, its world line is vertically upwards. If it is moving inertially with some velocity, then it has a straight world line which is inclined to the vertical. The faster the motion, the larger the angle with the vertical.

In reality, ordinary space has three dimensions and Newtonian space-time four. Instead of cards placed at vertical positions representing different times, or simultaneity levels, we must imagine three-dimensional spaces fused into a four-dimensional block. This is impossible to visualize, but the model with only two space dimensions is a good substitute.

Newtonian space-time differs in an important respect from space, in which all directions are on an equal footing and none is distinguished from any other. In Newtonian space-time, one direction is singled out. This is reflected in its representation as a pack of cards. Directions that lie in a card, in a simultaneity level, are quite different from the time line that runs vertically through the cards. Newtonian space-time is 'laminated'. If you were to 'cut through it' at an angle, the 'lamination' would be revealed. You would be 'cutting through' the simultaneity levels. The inequivalence of directions can be expressed in the language of coordinates.

Just as you can put a coordinate grid on a two-dimensional map, you can 'paint' a rectangular grid on Newtonian space-time with one of the axes perpendicular (parallel to the time line). The laws of motion can be formulated in terms of the grid. For example, bodies moving inertially travel along lines that are straight relative to the grid. You can then 'move' the grid around as a complete unit into different positions in space-time and see if the motions relative to the new grid satisfy the same laws as they did in the old. For Newton's laws there is considerable but not complete freedom to move the grid. Provided it is maintained vertical, it can be shifted and rotated in ordinary space, just like a child's climbing frame, and it can also be raised and lowered in the vertical time direction. However, tilting the vertical axis is not allowed. Newtonian forces (in gravity and electrostatics, for example) are transmitted instantaneously – horizontally in the model. If you tilt the grid from the original time axis, you leave the old simultaneity levels. The forces are not transmitted through the new levels.

Minkowski's real discovery was that, in an analogous construction using Maxwell's electromagnetic equations instead of Newton's laws, the resulting space-time structure, now called Minkowski space-time, has no special 'lamination'. It is more like a loaf of bread, through which you can slice in any way. The cut surface always looks the same. The way this shows up in changes of the coordinate grid is especially striking. Time becomes very like space but not quite identical.

The difference can be illustrated by the climbing frame. Here too a vertically held frame can be shifted, rotated and raised or lowered as a rigid unit. Maxwell's laws still take the same form with respect to the displaced grid. But you can also tilt it from the vertical provided you do something else as well. For this, you need an 'articulated' grid, which we have in fact already encountered, in Figure 25 in the discussion of simultaneity. It is a typical example of the space-time diagrams that Minkowski introduced. Figure 26, with its remarkable demonstration that two families of observers moving relative to each other each see the rods of the others as contracted relative to their own, is one of Minkowski's actual diagrams, slightly modified (merely to conform with the context of this book – the physical content remains unchanged).

In Figure 25, the original grid is 'painted' onto space-time with the dashes, while the dots show an alternative. As we saw, the law of nature that describes the behaviour of light pulses allows them to travel along the diagonals of either grid. A transformation of this law from one coordinate grid to another is called a *Lorentz transformation*, and the grids themselves are called *Lorentz frames*. I have already mentioned that you should not think in terms of there being one rectangular coordinate grid, and all the others oblique. Alice thinks Alice* has an oblique system compared to her, but Alice* thinks the same about Alice's system. This is a consequence of the relativity principle, and a special property of space-time that we shall come to shortly. Minkowski pointed out that the transformation shown in Figure 25 is a kind of rotation in four dimensions. The possibility of making rotations in ordinary space is a deep reflection of its unitary, block-like nature. Minkowski saw the possibility of making a kind of rotation in space-time, which is impossible in Newtonian space-time, as the clearest evidence for the intimate fusion of space and time, even though the need for 'articulation' showed that time was still somewhat different in nature from space.

Einstein, Minkowski and others were able to show that all the laws of nature known in their time (except initially for gravitation) either already had a form that was exactly the same in all Lorentz frames or could be relatively easily modified so that they did. Even though the modifications were relatively easy once the idea was clear, their implications, including Einstein's famous equation $E = mc^2$ (a prediction at that time), were mostly very startling. Minkowski, like Einstein and Poincaré, made a strong prediction that all laws of nature found in future would accord with the relativity principle, and emphasized that the guiding principle for finding such laws was to treat time exactly as if it were space.

Except for the intermingling of space and time and the distinguished role played by light, Minkowski's space-time strongly resembles Newtonian space-time. Matter neither creates nor changes its rigid and absolute structure. It is like a football field, complete with markings, on which the players must abide by rules they cannot change.

ARE THERE NOWS IN RELATIVITY?

It is often said that relativity destroyed the concept of Now. In Newtonian physics the axes can never be tilted as they are in Figure 25. The simultaneity levels stay level, and there is a unique sequence of instants of time, each of which applies to the complete universe. This is overthrown in relativity, where each event belongs to a multitude of Nows. This has important implications for the way we think about past, present and future.

Even in Newtonian theory we can picture world history laid out before us. In this 'God's-eye' view, the instants of time are all 'there' simultaneously. The alternative idea of a 'moving present' passing through the instants from the past to the future is theoretically possible but impossible to verify. It adds nothing to the scientific notion of time. Special relativity makes a 'moving present' pretty well untenable, even as a logical possibility.

Imagine that two philosophers meet on a walk. Each believes in a present that sweeps through instants of time. But that implies a unique succession of instants, or Nows. Which Nows are they? If the two philosophers are to make such claims, they should be able to 'produce' the Nows through which time flows. Unfortunately, they face the problem of the relativity of simultaneity. Each can define simultaneity relative to themselves, but, since they are walking towards each other, their Nows are different, and that puts paid to any idea that there is a unique flow of time. There is no natural way in which time can flow in Minkowski's space-time. At least within classical physics, space-time is a block – it simply is. This is known as the block universe view of time. Everything – past, present and future – is there at once. Some authors claim that nothing in relativity corresponds to the experienced Now: there are just point-like events in space-time and no extended Nows. At the psychological level, Einstein himself felt quite disturbed about this. Reporting a discussion, the philosopher Rudolf Carnap wrote:

> Einstein said the the problem of the Now worried him seriously. He explained that the experience of the Now means something special for man, something essentially different from the past and the future, but that this important difference does not and cannot occur within physics. That this experience cannot be grasped by science seemed to him a matter of painful but inevitable resignation. So he concluded 'that there is something essential about the Now which is just outside the realm of science'.

The block universe picture is in fact close to my own, but the idea that Nows have no role at all to play in physics, and must be replaced by point-like events, would destroy my programme. However, it is only absolute simultaneity that Einstein denied. Relative simultaneity was not overthrown.

We are all familiar with flat surfaces (two-dimensional planes) in three-dimensional space. Such planes have one dimension fewer than the space in which they are embedded, and are flat. *Hyperplanes* are to any four-dimensional space what planes are to space. In Newtonian physics, space at one instant of time is a three-dimensional hyperplane in four-dimensional Newtonian space-time. It is a *simultaneity hyperplane*: all points in it are at the same time. Such hyperplanes also exist in Minkowski space-time, but they no longer form a unique family. Each splitting of space-time into space and time gives a different sequence of them.

Now, what is Minkowski space-time made of? The standard answer is events, the points of four-dimensional space-time. But there is an alternative possibility in which three-dimensional configurations of extended matter are identified as the building blocks of space-time. The point is that the three-dimensional hyperplanes of relative simultaneity are vitally important structural features of Minkowski space-time. It is an important truth that special relativity is about the existence of distinguished frames of reference. And an essential fact about them is that they are 'painted' onto simultaneity hyperplanes. As a consequence, simultaneity hyperplanes, which are Nows as I define them, are the very basis of the theory. They are distinguished features. You cannot begin to talk about special relativity without first introducing them. At this point, the way both Einstein and Minkowski created special relativity becomes significant.

The question is this: how is a four-dimensional structure built up from three-dimensional elements? To make this easier to visualize, consider the analogous problem of building up a three-dimensional structure from cards with marks on them representing the distribution of matter. From one set of cards with given marks, many different structures can be built simply by sliding the cards horizontally relative to one another and changing their vertical spacings. Tait's problem shows that in general the markings in a structure built without special care will not satisfy the laws of motion. What is more, to find the correct positionings we have to use the complete extended matter distributions. These are what I have identified as instants of time. You simply cannot make the space-time structure without using them.

The interesting thing is that neither Einstein nor Minkowski gave serious thought to this problem – they simply supposed it had been solved. They started their considerations at the point at which space-time had already been put together. A comment by Minkowski, more explicit than Einstein, makes this clear: 'From the totality of natural phenomena it is possible, by successively enhanced approximations, to derive more and more exactly a system of reference x, y, z, t, space and time, by means of which these phenomena then present themselves in agreement with definite laws.' He then points out that one such reference system is by no means uniquely determined, and that there are transformations that lead from it to a whole family of others, in all of which the laws of nature take the same form. However, he never says what he means by 'the totality of natural phenomena' nor what steps must be taken to perform the envisaged successive approximations. But how is it done? This is a perfectly reasonable question to ask. We are told how to get from one reference system to another but not how to find the first one. Had either Einstein or Minkowski asked this question explicitly, and gone through the steps that must be taken, then the importance of extended matter configurations, and with them instants of time as I define them, would have become apparent. This is a key part of my argument. The accidents of the historical development have obscured the vital role of extended Nows and given the erroneous impression that events are primary.

I am not claiming that the description of space-time given by Einstein or Minkowski is wrong. Far from it – they got it right, but they described the finished product, and the complete story must also include the construction of the product. This is best done directly for the space-time of general relativity, the topic of the next chapters. As preparation for them, I conclude this chapter with a summary of the most important points.

Minkowski space-time is not some amorphous bulk in which there is no simultaneity structure at all. We can 'paint coordinate lines' – and an associated simultaneity structure – on space-time in many different ways. But the whole content of the theory would be lost if we could not do it one way or the other. There is no doubt about it – simultaneity hyperplanes exist out there in space-time as distinguished features.

Moreover, to give any content to relativity, we must, almost paradoxically, assume a universality of three-dimensional things. The clocks we can find in one Lorentz frame must be identical to the clocks we can find in any other. This is a prerequisite of the relativity principle, for it says that the laws of physics are identical in any such frame. That would be impossible if a particular kind of clock could exist in one frame but not in another. We can go further. On any hyperplane in any Lorentz frame, the actual things in the world (electric and magnetic fields, charged particles, etc.) can have any one of a huge number of different arrangements. Each of them is just like the possible distributions of particles from which we constructed Platonia for Newtonian physics.

Exactly the same thing can be done in relativity. There is a Minkowskian Platonia, whose points are all possible distributions of fields and matter that one can find on any simultaneity hyperplane in Minkowski space. Whatever Lorentz frame we choose, the Minkowskian Platonia always comes out the same. If it were not, the relativity principle, with its insistence that the laws of nature are identical in all Lorentz frames, would be meaningless. To be identical, the laws must operate on identical things, which are precisely the distributions that define the points of Platonia. For all its four-dimensional integrity, space-time is built of three-dimensional bricks. The beautiful four-dimensional symmetries hide the vital role of the bricks.

It is just that space-time is not constructed from a unique set. The analogy with a pack of cards is again quite apt. Newtonian space-time is an ordinary pack; Minkowski space-time is a magical pack. Look at it one way, and cards run through the block with one inclination. Look at it another way, and different cards run with a different inclination. But whichever way you look, cards are there.

CHAPTER 10

The Discovery of General Relativity

FUNNY GEOMETRY

This chapter is about how Einstein progressed from special relativity, which does not incorporate gravity, to general relativity, which does. Einstein believed that he was simultaneously incorporating Mach's principle as its deepest foundation, but later, as I said, he changed his mind and left this topic in a great muddle. My view is that, nevertheless, without being aware of it, Einstein did incorporate the principle. This has important implications for time. We start with a bit more about Minkowski's discoveries, which is necessary if we are to understand the way Einstein set about things.

One of the most important concepts in physics and geometry is distance, which is measured with rods. Distances can be measured in a space of any number of dimensions. You can measure them along a line or curve, on a flat or curved surface, or in space. In Part 2 we saw how an abstract 'distance', the action, can be introduced in multidimensional configuration spaces like Platonia. Minkowski showed that a remarkable kind of four-dimensional distance exists in space-time. Its existence is a consequence of the experimental facts that underlie special relativity. These things are most easily explained if we assume that space has just one dimension, not three; space-time then has two dimensions. Such a space-time is shown in Figure 27. We must first of all learn about past, present, and future in space-time.

One of the distinguished coordinate systems that exists in space-time is shown in Figure 27, in which the x axis is for space and the t axis for time, which increases upward. This is the Lorentz frame of Alice in Figure 25.

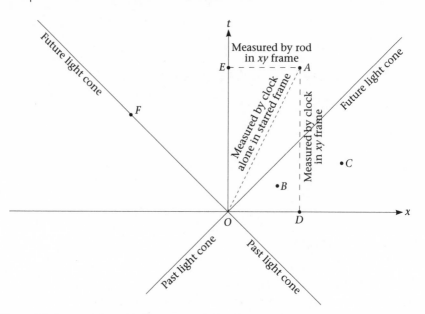

Figure 27 Past and future light cones and the division of space-time in time-like and space-like regions, as described in the text.

Her world line is the vertical t axis. The units of time and distance are chosen to make the speed of light unity. Light pulses that pass through event O at $t = 0$ in opposite directions in space travel in space-time along the two lines marked *future light cone*. Their continuations backward (the light's motion before it reaches O) define the *past light cone*.

Each event has a light cone, but only O's is shown. Relativity differs from Newtonian theory mainly through the light cone and its associated distinguished speed c, which is a limiting speed for all processes. Light plays a distinguished role in relativity simply because it has that speed. No material object can travel at or faster than it. If a material object passes through O, its world line must lie somewhere inside the light cone, for example OA in Figure 27.

The light cone divides space-time into qualitatively different regions. An event like A can be reached from O by a material object travelling slower than light. Two such events are *time-like* with respect to each other. For two such events there exists a Lorentz frame in which they have the same space coordinates but different time coordinates. For the points O and A this frame is shown in the upper right of Figure 28.

Next we consider events like B and C in Figure 27, outside the light

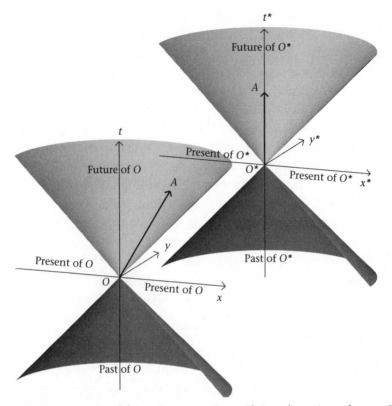

Figure 28 Past, present and future in a space-time with two dimensions of space. The object that moves along *OA* (bottom left) is at rest in the starred frame (top right). Its world line is *O*A* (*O* and *O** are the same event).

cone of *O*. They are *space-like* with respect to *O*. No material body can reach them from *O*, since to do so it would have to travel faster than light. For two events that are mutually space-like there exists a Lorentz frame in which they have the same time coordinate but different space coordinates. For two space-like events, it is impossible to say which is the earlier in any absolute sense. In some Lorentz frames one will be earlier than the other (thus *O* is earlier than both *B* and *C* in Alice's frame in Figure 27), but in others the temporal order will be reversed.

Finally, two events that can be connected by a light ray have a *light-like* relationship. All points on the light cone of event *O*, for example the point *F*, are light-like with respect to *O*.

These three basic relationships between events – being time-like, space-like or light-like – are the same in all Lorentz frames. This is because the three types are determined by the light cones, which are real features in space-time, just as rivers are real features of a continent. In contrast, the coordinate axes are like lines 'painted' on space-time – they are no more real than the grid lines on a map. Moreover, in a change from one frame to another, the coordinate axes never cross the light cones. The time axis moves but stays within the light cone, while the space axes stay within the 'present' as defined above. This is illustrated in Figure 28 for space with two dimensions, which shows how the light cone gets its name. It also highlights the great difference between the Newtonian and Einsteinian worlds. In the former, past, present and future are defined throughout the universe, and the present is a single simultaneity hyper-plane. In the latter, they are defined separately for each event in space-time, and the present is much larger.

Now we can talk about distance. In ordinary space it is always positive. The distance relationships are reflected in Pythagoras' theorem: the square of the hypotenuse in any right-angled triangle is equal to the sum of the squares of the other two sides: $H^2 = A^2 + B^2$.

Minkowski was led to introduce a 'distance' in space-time by noting a curious fact. For observers who use the xy frame in Figures 27 and 28, event A is separated from O by the space-like interval EA and by the time-like interval DA. For observers who use the starred frame, however, O and A are at the same space point and are merely separated by the time-like interval OA. The xy observers measure EA with a rod and DA with a clock, obtaining results X and T, respectively. With their clock, observers in the starred frame can measure only the time-like interval OA. Now, their clock runs at a different rate to the xy clock, so they will find that OA is not T but $T_{starred}$. Using Einstein's results, Minkowski found that $(T_{starred})^2 = T^2 - X^2$. This is just like Pythagoras' theorem, except for the minus sign.

There are several important things about this result. Einstein had shown that observers moving relative to each other would not agree about distances and times between pairs of events. However, Minkowski found something on which they will always agree. Measurements of the space-like separation (by a rod) and the time-like separation (by a clock) of the same two events O and A can be made by observers moving at any speed. They will all disagree about the results of the separate measure-ments, but they will all find the same value for the square of the time-like separation *minus* the square of the space-like separation. It will always be equal to the square of the time-like separation, called the *proper time*, of the unique observer for whom *O* and *A* are at the same space position. This result created a sensation. Space and time, like rods and clocks, seem to have completely different natures, but Einstein and Minkowski showed that they are inseparably linked.

What is more, Minkowski showed that it is very natural to regard space and time together as a kind of four-dimensional country in which any two points (events in space-time) are separated by a 'distance'. This 'distance', found by measurements with both rods and clocks, is to be regarded as perfectly real because everyone will agree on its value. In fact, Minkowski argued that it is more real than ordinary distances or times, since different observers disagree on them. Only the 'distance' in space-time is always found to be the same. But it is a novel distance – positive for the time-like *OA* in Figure 27, zero for the light-like *OF* and negative for the space-like *OC*. (It is a convention, often reversed, to make time-like separations positive and space-like ones negative. What counts is that they have opposite signs. Also, if the units of space and time are not chosen to make the speed of light *c* equal to 1, the square of the space-time 'distance' becomes $(cT)^2 - X^2$.)

Almost everything mysterious and exciting about special relativity arises from the enigmatic minus sign in the space-time 'distance'. It causes the 'skewing' of both axes of the starred frame of the starred twins in Figure 25, and leads to the single most startling prediction – that it is possible, in a real sense, to travel into the future, or at least into the future of someone else, since the future as such is not uniquely defined in special relativity. What we call space and time simply result from the way observers choose to 'paint coordinate systems' on space-time, which is the true reality. Minkowski's diagrams made all these mysteries transparent – and intoxicatingly exciting for physicists. However, this is not the place to discuss time travel and the other surprises of relativity, which are dealt with extensively in innumerable other books.

EINSTEIN'S WAY TO GENERAL RELATIVITY

For physicists, 'relativity' has two different meanings. The more common is the one employed by Einstein when he created relativity. He related it to the empirical fact, first clearly noted by Galileo in 1632, that all observations made within an enclosed cabin on a ship sailing with uniform speed are identical to observations made when the ship is at rest. Einstein illustrated this fact with experiments on trains. The lesson he drew from it was that uniform motion as such could not be detected by any experiment. The laws of nature could therefore not be expressed in a unique frame of reference known to be at rest. They could be expressed only in any one of a family of distinguished frames in uniform motion relative to one another. The *relativity principle* states that the laws of nature have the identical form in all such frames. For reasons shortly to be explained, this later became known as the *restricted* or *special* relativity principle.

This meaning of relativity is tied to a special feature of the world – the existence of the distinguished frames and their equivalence for expressing the laws of nature. The other meaning of relativity is more primitive and less specific. It simply recognizes that space and time are invisible: all we ever see are objects and their relative motions. We can speak meaningfully of the position and motion of an object only if we say how far it is from other objects. Position and motion are relative to other objects. This is often called *kinematic relativity*, to distinguish it from Galilean relativity.

Both relativity principles have played important – often decisive – roles in physics. Copernicus and Kepler used kinematic relativity to great effect in the revolution they brought about. Galileo used the other relativity principle to explain how we can live on the Earth without feeling its motion. That was almost as wonderful a piece of work as Einstein's, nearly three hundred years later. A natural question is this: what is the connection between the two relativity principles? Any satisfactory answer must grapple with and resolve the issue of the distinguished frames of reference. How are they determined? What is their origin? As we have seen, neither Einstein nor Minkowski addressed these questions when they created special relativity, and they have been curiously neglected ever since. This is a pity, since they touch upon the nature of time. We cannot say what time is – and whether it even exists – until we know what motion is.

Poincaré sought to unite the two relativity principles in a single condition on the structure of dynamics, as formulated in the two-snapshots idea. Had he succeeded, he would have derived the empirical fact of Galilean relativity solely on the basis of a natural criterion derived from kinematic relativity. He died without taking this idea any further, but in any case it is doubtful whether the two relativity principles can be fully fused into one. Poincaré formulated his idea in 1902, before the relativistic intermingling of space and time became apparent, and it is hard to see how that can ever be derived from the bare fact of kinematic relativity. It is, however, of great interest to see how far Poincaré's idea can be taken. We shall come to this when we have seen how Einstein thought about and developed his own relativity principle and thereby created general relativity.

It is important not to be overawed by the genius of Einstein. He did have blind spots. One was his lack of concern about the determination in practice of the distinguished frames that play such a vital role in special relativity – he simply took them for granted. It is true that they are realized approximately on the reassuringly solid Earth in skilfully engineered railway carriages. But how does one find them in the vast reaches of space? This is not a trivial question. Matching this lack of practical interest, we find an absence of theoretical concern. Einstein asked only what the laws of nature look like in given frames of reference. He never asked himself whether there are laws that determine the frames themselves. At best, he sought an indirect answer and got into a muddle – but a most creative muddle.

To see why, it is helpful to trace the development of his thinking – a fascinating story in its own right. As an extremely ambitious student, he read Mach's critique of Newton's absolute space. This made him very sceptical about its existence. Simultaneously, he was exposed to all the issues related to the aether in electrodynamics. Lorentz, in particular, had effectively identified absolute space with the aether, in the form of an unambiguous state of rest. But, writing to his future wife Mileva in August 1899, Einstein was already questioning whether motion relative to the aether had any physical meaning. This would develop into one of the key ideas of special relativity. If it is impossible to detect motion relative to it, the aether cannot exist. It was natural for Einstein to apply the same thought to absolute space.

His 1905 paper killed the idea that uniform motion relative to any kind of absolute space or aether could be detected. But Newton had based his case for absolute space on the detection not of uniform motion, but of acceleration. In 1933, Einstein admitted that in 1905 he had wanted to extend the relativity principle to accelerated as well as uniform motion, but could not see how to. The great inspiration – 'the happiest thought of my life' – came in 1907 when he started to consider how Newtonian gravity might be adapted to the framework of special relativity. He suddenly realized the potential significance of the fact, noted by Galileo and confirmed with impressive accuracy by Newton, that all bodies fall with exactly the same acceleration in a gravitational field.

Most physicists saw this as a quirk of nature, but Einstein immediately decided to elevate it to another great principle and exploit it as he had the relativity principle. Unable to divine new laws of gravitation straight off, he formulated the *equivalence principle*, according to which processes must unfold in a uniform gravitational field in exactly the same way as in a frame of reference accelerated uniformly in a space free of gravity. He argued that pure acceleration could not be distinguished from uniform gravitation. Suppose that you awoke from a deep narcotic sleep in a dark bedroom to find that gravity was mysteriously stronger. There could be two different causes. You might have been transported, bedroom and all, to another planet with stronger gravity. But you might still be on the Earth but in an elevator accelerating uniformly upward. No experiments you could perform in your bedroom would enable you to distinguish between these alternatives.

Einstein saw here a striking parallel with the relativity principle. The relativity principle prevented an observer from detecting uniform motion. In its turn, the equivalence principle prevented an observer from detecting uniform acceleration – observed acceleration could be attributed either to acceleration in gravity-free space or to a gravitational field. Einstein recognized the immediate short-term potential of his new principle. He knew how processes unfolded in gravity-free space. Mere mathematics showed how they would appear in an accelerated frame, but by the equivalence principle it was possible to deduce that these same processes must occur in a uniform gravitational field. Once again, Einstein's inspired selection of a simple universal principle – all bodies fall in the same way – enabled him to perform a startling conjuring trick. He showed that the rate of clocks must depend on their position in a gravitational field. Clocks closer to gravitating bodies must run slow relative to clocks farther away.

This fact is often said to show that 'time passes more slowly' near a gravitating body. However, objective facts within relativity can seem utterly mysterious and logically impossible if we imagine time as a river. Such a time does not exist. Relativity makes statements about actual clocks, not time in the abstract. It is easy to imagine – and physicists now find it comparatively easy to verify – that otherwise identical clocks run at different rates at the top and bottom of a higher tower. Incidentally, the 'time dilation' effect in gravity is much easier to accept than the similar effect associated with motion. There is no reciprocal slowing down. Thus, observers at the top and bottom of the tower both agree that the clock at the top runs faster.

By 1907 Einstein was also able to show that gravity must deflect light. Both his early predictions, made precise in his fully developed theory, have been confirmed with most impressive accuracy in recent decades. However, Einstein saw his early predictions merely as stepping stones to something far grander. The equivalence principle persuaded him that inertia (i.e. the tendency of bodies to persist in a state of rest or uniform motion) and gravity, which Newton and all other physicists had regarded as distinct, must actually be identical in nature. He started to look for a conceptual framework in which to locate this conviction. At the same time, he saw a great opportunity to abolish not only the aether but also all vestiges of absolute space. So far he had managed to achieve two steps in this process by showing that uniform motion and uniform acceleration could not correspond to anything physically real in the world. However, much more general motions could be imagined. Einstein aimed to show that the laws of nature could be expressed in identical form whatever the motion of the frame of reference.

The relativity he had so far established was very special. What he wanted was complete *general relativity*. This idea, nurtured and developed over eight years and involving intense and often agonizing work during the last four, explains the name he gave to his unified theory of gravitation and inertia that finally emerged in 1915. Viewed in the light of the ancient debate about absolute and relative motion, Einstein's approach was very distinctive and somewhat surprising since he made no attempt to build kinematic relativity directly into the foundations of his theory. Unlike Mach and many other contemporaries, he did not insist that only relative quantities should appear in dynamics. He went at things in a roundabout way, mostly because of his preference for general principles. However, I think it was also a result of the way he thought about space and time.

As far as I can make out, Einstein did conceive of space-time as real and as the container of material things – fields and particles. However, he recognized that all its points were invisible and that they could be distinguished and identified only by observable matter present at them. Since space-time was made 'visible' by such matter, he supposed he could lay out coordinate grid lines on space-time and express the laws of nature with respect to them.

Now came the decisive issue. Einstein saw space-time without any matter in it as a blank canvas. Nothing about it could suggest why the coordinate grid lines should be drawn in one way rather than another. Any choice would be arbitrary and violate the principle of sufficient reason. Einstein found this intolerable. That is no exaggeration: his faith in rationality of nature – as opposed to human beings – was intense. The only satisfactory resolution was general relativity. In truth, there can be no distinguished coordinate systems. It must be possible to express the laws of nature in all systems in exactly the same form.

The only justification for the distinguished systems that appeared in Newtonian dynamics and special relativity was the law of inertia. But the equivalence principle had opened up the possibility of unifying inertia and gravity. This insight sustained Einstein in his long search for general relativity. His contemporaries would all have been content simply to find a new law of gravity. He was after something sublime.

It is suggestive that both Poincaré and Einstein – the old and young giants – began their attack on absolute space from the principle of sufficient reason. The difference between their approaches is interesting. Working within the traditional dynamical framework, Poincaré said that only directly observable quantities – the relative separations of bodies and their rates of change – should be allowed as initial data for dynamics. In such a theory, we may say that *perfect Laplacian determinism* holds (it doesn't hold in Newtonian theory, which uses invisible absolute space and time). Einstein had a more general approach. He merely insisted that there should be no arbitrary choice of the coordinate systems used to express the laws of nature.

THE MAIN ADVANCES

The desire to express the laws of nature in progressively more general coordinates led to all Einstein's major breakthroughs. Newton had argued that centrifugal forces proved the existence of absolute space. The laws of nature looked different in rotating systems. Einstein wanted to attack this problem head on. Could he perhaps show that, if expressed properly, the laws of nature did after all have the same form in rotating and non-rotating coordinates? The principle of equivalence suggested that what Newton had taken to be absolute inertial effects in a rotating system might be the gravitational effects of distant matter. The point is that in a rotating system the distant stars would themselves appear to be rotating. Since rotating electric charges generate electric and magnetic fields, rotating masses might generate new kinds of gravitational field. Nearly thirty years earlier, Mach had suggested that rotating matter 'many leagues thick' might generate measurable centrifugal forces within it. Einstein now conjectured that the gravitational field was the mechanism through which such forces could arise.

He therefore started to consider what form the laws of nature would take in a rotating system. This immediately led him to a startling conclusion: the ordinary laws of Euclidean geometry could not hold in such a system! His argument was based on the contraction of measuring rods in motion which he had proved in special relativity. First, imagine observers at rest on a surface who measure the circumference and diameter of a circle painted on it. They will find that their ratio is π. That agrees with Euclidean geometry – a recognized law of nature. Now imagine other observers on a disk above the painted circle and rotating about its centre. Their rods will undergo Fitzgerald–Lorentz contraction when laid out in the direction of motion, around the circumference. However, when laid out along the diameter, the rods will not contract. (The contraction occurs only in the direction of motion.) Therefore, the rotating observers will not find π when they measure the ratio of the circumference to the diameter. For them, Euclidean geometry will not hold.

Because Einstein wanted so passionately to generalize the relativity principle, he took this result seriously. According to the hint from the equivalence principle, novel effects in accelerated coordinate systems (as a rotating one is) could be attributed to gravitational effects. He concluded that geometry would not be Euclidean in a gravitational field. This happened during 1911/12, when he was working in Prague. Through either the suggestion of a colleague or the recollection of lectures on non-Euclidean geometry he had heard as a student, Einstein's attention was drawn to a classic study in the 1820s by the German mathematician Carl Friedrich Gauss.

Gauss had studied the curvature of surfaces in Euclidean space. As a rule, material surfaces in space are not flat but curved. Think of the surface of the Earth or any human body. Gauss's most important insight was that a surface in three-dimensional space is characterized by two distinct yet not entirely independent kinds of curvature. He called them *intrinsic* and *extrinsic* curvature. The intrinsic curvature depends solely on the distance relationships that hold within the surface, whereas the extrinsic curvature measures the bending of the surface in space. A surface can be flat in itself – with no intrinsic curvature – but still be bent in space and therefore have extrinsic curvature. The best illustration of this is provided by a flat piece of paper, which has no intrinsic curvature. As it lies on a desk it has no extrinsic curvature either: it is not bent in space. However, it can be rolled into a tube. It is then bent – but not stretched – and acquires extrinsic curvature.

In contrast to a sheet of paper, the surface of a sphere, like the earth, has genuine intrinsic curvature. Gauss realized that important information about it could be deduced from distance measurements made entirely within the surface. Imagine that you can pace distances very accurately, and that you walk due south from the north pole until you reach latitude 85° north. Then you turn left and walk due east all the way round the Earth at that latitude. All the time you will have remained the same distance R from the north pole. If you believed the Earth to be flat, you would expect to have to walk the distance $2\pi R$ before returning to the point of your left turn. However, you will find that you get there having walked a somewhat shorter distance. This shows you that the surface of the Earth is curved.

To describe these things mathematically for all smooth surfaces, Gauss found it convenient to imagine 'painting' curved coordinate lines on the surface. On a flat surface it is possible to introduce rectangular coordinate grids, but not if the surface is curved in an arbitrary way. So Gauss did the next best thing, which is to allow the coordinate lines to be curved, like the lines of latitude and longitude on the surface of the Earth. He showed how the distance between any two neighbouring points on a curved surface could be expressed by means of the distances along coordinate lines, and also how exactly the same distance relations could be expressed by means of a different system of coordinates on the same surface. About thirty years later, another great German mathematician, Bernhard Riemann, showed that not only two-dimensional surfaces but also three-dimensional and even higher-dimensional spaces can have intrinsic curvature. This is hard to visualize, but mathematically it is perfectly possible. Just as on the Earth, in a curved space of higher dimensions, you can, travelling always in the same direction, come back to the point you started from. These more general spaces with curvature are now called Riemannian spaces.

Einstein realized that he had to learn about all this work thoroughly, and it was very fortunate that he moved at that time to Zurich, where Marcel Grossmann, an old friend from student days, was working. Grossmann gave him a crash course in all the mathematics he needed. When he had fully familiarized himself with it, Einstein became extremely excited for two reasons.

First, Minkowski had shown that space-time could be regarded as a four-dimensional space with a 'distance' defined in it between any two points. Except that the 'distance' was sometimes positive and sometimes negative, whereas Riemann had assumed the distance to be always positive and had never envisaged time as a dimension, considered mathematically Minkowski's space-time was just like one of Riemann's spaces. But it was special in lacking curvature – it was like a sheet of paper rather than the Earth's surface. Einstein had meanwhile become convinced that gravity curves space-time. This led to one of his most beautiful ideas: in special relativity, the world line (path) of a body moving inertially is a straight line in space-time. This is a special example of a 'shortest curve', or geodesic. The corresponding path in a space with curvature would be a geodesic, like a great circle on a sphere.

Einstein assumed that the world line of a body subject to inertia *and gravity* would be a geodesic. In this way he could achieve his dream of showing that inertia and gravity were simply different manifestations of the same thing – an innate tendency to follow a shortest path. This will be a straight line if no gravity is present, so that space-time has no curvature, but in general it will be a curved (but 'straightest') line in a genuinely curved space-time. Since matter causes gravity, Einstein assumed that matter must curve space-time in accordance with some law, for which he immediately started to look. Bodies moving in such a space-time would follow the geodesics corresponding to the curvature produced by the matter, so the gravitational effect of the matter would be expressed through the curvature it produces. Another important insight was that in small regions the effect of curvature would be barely noticeable, just as the Earth seems flat in a small region, so that in those small regions physical phenomena would appear to unfold just as in special relativity without gravity. This gave full expression to the equivalence principle.

The second reason why Einstein became so excited was that Gauss's method matched his own idea of general relativity. He disliked the distinguished frames of special relativity because they corresponded to special ways of 'painting' coordinate systems onto space-time. He felt that this was the same as having absolute space and time. They would be eliminated only if the coordinate systems could be painted on space-time in an arbitrary way. But this was what Gauss's method amounted to. In fact, in a curved space it is mathematically impossible to introduce rectangular coordinates. Mathematicians call the possibility of using completely arbitrary coordinate systems *general covariance*. Specifically, laws are said to be generally covariant if they take exactly the same form in all coordinate systems. Einstein identified this with his requirement of general relativity.

To summarize this part of the story, in 1912 Einstein became aware of the possibilities opened up by non-Euclidean geometry and the work initiated by Gauss. He had begun to suspect that gravitational fields would make geometry non-Euclidean. He was also almost desperate to find a formalism that did not presuppose distinguished frames of reference. He found that Gauss's method of arbitrary coordinates was tailor-made for his ambitions. He also saw that, space and time having been so thoroughly fused by Minkowski, the only natural thing to do was to make space-time into a kind of Riemannian space. The ideas of Gauss and Riemann must be applied, not to space alone, but to space and time. This is the incredibly beautiful idea that Minkowski made possible: gravity was to be explained by curvature in space *and time*. Einstein thus conjectured that space-time is curved by gravity, and that bodies subject only to gravity and inertia follow geodesics determined by the distance properties of space-time, which encapsulate all its geometrical properties. Einstein's conjecture has been brilliantly confirmed to great accuracy in recent decades.

THE FINAL HURDLE

Finding the law of motion of bodies in a gravitational field was only part of Einstein's problem. He also had to find how matter created a gravitational field. He needed to find equations for the gravitational field somewhat like those that Maxwell had found for the electromagnetic field. They would establish how matter interacted with the gravitational field, and also how the field itself varied in regions of space-time free of matter (matching the way electromagnetic radiation propagated as light through space-time). This part of the problem created immense difficulties for Einstein, mostly through very bad luck.

Much as I would like to tell the complete story, which is fascinating and full of ironies, I shall have to content myself with saying that, after three nerve-wracking years, Einstein finally found a generally covariant law that described how matter determined the curvature of space-time. It involves mathematical structures called tensors, all the properties of which had already been studied by mathematicians. In particular, for space-time free of matter, Einstein was able to show that a tensor known as the Ricci tensor (because it had been studied by the Italian mathematician Gregorio Ricci-Curbastro) must be equal to zero. Ironically, Grossmann had already suggested to Einstein in 1912 that in empty space the vanishing of the Ricci tensor might be the generally covariant law he was seeking. However, some understandable mistakes prevented them from recognizing the truth at that time.

It is a striking fact that all the mathematics Einstein needed already existed. In fact, I believe it is significant that he did not have to invent any of it. In 1915, he was immediately able to show that, to the best accuracy astronomers could achieve at that time, his theory gave identical predictions to Newtonian gravity except for a very small correction to the motion of Mercury. All planetary orbits are ellipses. A planet's elliptical orbit itself very slowly rotates, under the gravitational influence of the other planets. This is known as the advance of the perihelion, the perihelion being the point at which the planet is closest to the Sun, marking one end of the ellipse's longest diameter. According to Einstein's theory, Mercury's perihelion should advance by 43 seconds of arc per century more than was predicted by Newtonian theory. This very small effect shows up for Mercury because it is closer to the Sun than the other planets, and also has a large orbital eccentricity. For many years, the sole discrepancy in the observed motions of the planets had been precisely such a perihelion advance for Mercury of exactly that magnitude. All attempts to explain it had hitherto failed. Einstein's theory explained it straight off.

GENERAL RELATIVITY AND TIME

Many more things could be said about general relativity and its discovery. However, what I want to do now is identify the aspects of the theory and the manner of its discovery that have the most bearing on time.

First, the classical (non-quantum) theory as it stands seems to make nonsense of my claim that time does not exist. The space-time of general relativity really is just like a curved surface except that it has four and not two dimensions. A two-dimensional surface you can literally see: it is a thing extended in two dimensions. In their mind's eye, mathematicians can see four-dimensional space-time, one dimension of which is time, just as clearly. It is true that time-like directions differ in some respects from space-like directions, but that no more undermines the reality of the time dimension than the difference between the east–west and north–south directions on the rotating Earth makes latitude less real than longitude. However, the qualification 'as it stands' at the start of this paragraph is important. In the next chapter we shall see that there is an alternative, timeless interpretation of general relativity.

Next, there is the matter of the distinguished coordinate systems. In one sense, Einstein did abolish them. Picture yourself in some beautiful countryside with many varied topographic features. They are the things that guide your eye as you survey the scene. The real features in space-time are made of curvature, and hills and valleys are very good analogies of them. Imagined grid lines are quite alien to such a landscape. In general relativity, the coordinated lines truly are merely 'painted' onto an underlying reality, and the coordinates themselves are nothing but names by which to identify the points of space-time.

For all that, space-time does have a special, sinewy structure that needs to be taken into account. Distinguished coordinate systems still feature in the theory. This is because the theory of measurement and the connection between theory and experiment is very largely taken over from special relativity. In fact, much of the content of general relativity is contained in the meaning of the 'distance' that exists in space-time. This is where the analogy between space-time and a landscape is misleading. We can imagine wandering around in a landscape with a ruler in our pocket. Whenever we want to measure some distance, we just fish out the ruler and apply it to the chosen interval. But measurement in special relativity is a much more subtle and sophisticated business than that. In general, we need both a rod and a clock to measure an interval in space-time. Both must be moving inertially in one of the frames of reference distinguished by that theory, otherwise the measurements mean nothing. The theory of measurement in general relativity simply repeats in small regions of space-time what is done in the whole of Minkowski space-time in special relativity. No measurements can be contemplated in general relativity until the special structure of distinguished frames that is the basis of special relativity has been identified in the small region in which the measurements are to be made.

This is something that is often not appreciated, even by experts. It comes about largely because of the historical circumstances of the discovery of general relativity and the absence of an explicit theory of rods and clocks. There is also the stability of our environment on the Earth and the ready availability in our age of clocks. It is easy for us to stand at rest on the Earth, watch in hand, and perform a measurement of a purely time-like distance. But nature has given us the inertial frame of reference for nothing, and skilful engineers made the watch. Finally, because we can and very often do see a three-dimensional landscape spread out before our eyes, it is very easy to imagine four-dimensional space-time displayed in the same way. All textbooks and popular accounts of the subject positively encourage us to do so. They all contain 'pictures' of space-time. Now the picture is indeed there, and very wonderful it is too. But it arises in an immensely sophisticated manner hidden away within the mathematical structure of the Ricci tensor. The story of time as it is told by general relativity unfolds within the Ricci tensor. It performs the miracle – the construction of the cathedral of space-time by intricate laying and interweaving of the bricks of time. I shall try to explain this in qualitative terms in the next chapter. Let me conclude this one by highlighting again the importance of the historical development. It made possible the discovery of a theory without full appreciation of its content.

At the end of November 1915, Einstein wrote an ecstatic letter to his lifelong friend Michele Besso, telling him that his wildest dreams had come true: 'General covariance. Mercury's perihelion with wonderful accuracy.' These two verbless sentences say it all. Einstein was convinced that general covariance had deep physical consequences and had led him to one of the greatest triumphs of all time. Yet, barely two and a half years later, he admitted, in response to a quite penetrating criticism from a mathematician called Erich Kretschmann, that general covariance had no physical significance at all.

In a way, this is obvious. Space-time is a beautiful sculpture. What makes it beautiful is the way in which its parts are put together. The fact that one can paint coordinate lines on the finished product and measure distance on the sculpture between points on it labelled by the arbitrary coordinates clearly leaves the sculpture exactly the same. All this changing of coordinates is purely formal. It tells you nothing about the true rules that make the sculpture.

Belatedly, Einstein came to see that his whole drive to achieve general covariance as a deep physical principle had no foundation in fact. It was just a formal mathematical necessity. Ever determined to find new and even more beautiful laws of nature, he never felt the need to go back and see exactly how his sculpture was actually created. In a book I wrote some years ago on the discovery of dynamics, I commented on the fact that Kepler (so very like Einstein in his dogged holding on to an idea that eventually transformed physics) never realized quite what a wonderful discovery he had made. I likened him to

> a boy who finds for the first time a ripe horse-chestnut with the outer shell intact. Cherishing the golden and curiously shaped object, he might take it home, quite unaware of the shiny brown and perfectly smooth conker ready to spring from the shell on application of a little directed pressure. That was Kepler's fate: he died without an inkling of what his nut really contained.

The same thing happened to Einstein. Realizing while still at Prague the sort of thing he needed, he hurried to a shop called 'Mathematics' owned by his friend Grossmann in Zurich. Straight off the shelf, at a bargain price, he bought a wonderful device called the Ricci tensor. Three years later, after agonizing struggles, he learned how to turn the handles properly, and out popped the advance of Mercury's perihelion and the exact light deflection at eclipses.

But it never entered his head to ask how the device actually worked. He died only half aware of the miracle he had created.

CHAPTER 11

General Relativity: The Timeless Picture

THE GOLDEN AGE OF GENERAL RELATIVITY

Strange as it may seem, general relativity was little studied for about forty years. This was not for want of admiration, for it was soon recognized as a supreme achievement. Confirmation of the predicted bending of starlight near the Sun by Arthur Eddington's eclipse expedition in 1919, communicated by telegram to *The Times*, made Einstein into a world celebrity overnight. The problem was that there seemed to be little one could do except wonder at the miracle of the theory he had created.

The main difficulty was the extreme weakness of all readily accessible gravitational fields. Apart from three small differences from Newtonian theory, which were all reasonably well confirmed, no further experimental tests seemed possible. A further problem was the mathematical complexity of the theory. Its solutions contained fascinating structures, above all black holes, but it was decades before these were discovered and fully understood. Finally, interest in general relativity was overshadowed by the discovery in 1925/6 of quantum mechanics. In fact, truly active research in general relativity commenced only in 1955, ironically the year Einstein died, with a conference held in Bern (where Einstein had worked as a patent clerk in 1905) to mark the fiftieth anniversary of special relativity.

Since then, research has concentrated in three main fields. First, there have been tremendous experimental advances, made possible above all by technological developments, including space exploration. The foundations and some detailed predictions of the theory have been tested to a very high accuracy. Particularly important was the discovery a quarter of a century ago of the first binary pulsar, observations of which have provided strong evidence for the existence of the gravitational waves predicted by the theory. General relativity has also played a crucial role in observational astronomy and cosmology.

There have been two broad avenues of theoretical research. First, general relativity has been studied as a classical four-dimensional geometrical theory of space-time, a systematical and beautiful development of Minkowski's pioneering work. Roger Penrose has probably done more than anyone else in this field, though many others, including Stephen Hawking, have made very important contributions. Second, the desire to understand the connection between general relativity and quantum mechanics (Box 2) has stimulated much work. Here it is necessary to distinguish two programmes. The less ambitious one accepts space-time as a classical background and seeks to establish how quantum fields behave in it. This work culminated in the amazing discovery by Hawking that black holes have a temperature and emit radiation. In *Black Holes and Time Warps*, Kip Thorne has given a gripping account of this story. Although the full significance of Hawking's discovery is still far from understood, nobody doubts its importance for the more ambitious programme, which is to transform general relativity itself into a quantum theory (Box 2). This transformation, which has not yet been achieved, is called the *quantization* of general relativity.

In fact, many researchers believe that it is a mistake to try to quantize general relativity directly before gravity has been unified with the other forces of nature. This they hope to achieve through superstring theory. However, a substantial minority believe that general relativity contains fundamental features likely to survive in any future theory, and that a direct attempt at its quantization is therefore warranted. This is my standpoint. In particular, I regard general relativity as a classical theory of time. It must surely be worth trying to establish its quantum form. Even if we have to await a future theory for the final details, the quantization of general relativity should give us important hints about the quantum theory of time.

It was the desire to quantize general relativity that led to the work described in this chapter. One important approach, called *canonical quantization*, is based on analysis of the dynamical structure of the classical theory. This is how general relativity came to be studied in detail as a dynamical theory nearly half a century after its creation as a geometrical space-time theory. The 'hidden dynamical core', or deep structure, of the theory was revealed. The decisive analysis was made in the late 1950s by Paul Dirac and the American physicists Richard Arnowitt, Stanley Deser and Charles Misner. They created a particularly elegant theory, now known universally as the ADM formalism. (Because it is regarded as controversial by some, the initials are occasionally reshuffled as MAD or DAM.)

The dynamical form of general relativity is often called *geometrodynamics*. The term, like 'black hole' and several others, was coined by John Wheeler, who, together with his many students at Princeton, did much to popularize this form of the theory. The interpretation of it proposed in this chapter is very close to one put forward by Wheeler in the early 1960s. However, I believe it brings out the essentially timeless nature of general relativity rather more strongly than Wheeler's well-known writings of that period. What is at stake here is the plan of general relativity. What are its ultimate elements when it is considered as a dynamical theory, and how are they put together?

This is what Dirac and ADM set out to establish. The answer was manifestly a surprise for Dirac at least, since it led him to make the remarkable statement quoted in the Preface. They found that if general relativity is to be cast into a dynamical form, then the 'thing that changes' is not, as people had instinctively assumed, the four-dimensional distances within space-time, but the distances within three-dimensional spaces nested in space-time. The dynamics of general relativity is about three-dimensional things: Riemannian spaces.

PLATONIA FOR RELATIVITY

To connect this with the topics of Part 2, let me tell you about the work that Bruno Bertotti and I did after the work described there. We began to wonder whether we could be more ambitious and construct not merely a non-relativistic, Machian mechanics, but perhaps an alternative to general relativity. At the time, we believed that Einstein's theory did not accord with genuine Machian principles. Experimental support for it was beginning to seem rather convincing, but tiny effects have often led to the replacement of a seemingly perfect theory by another with a very different structure. We were aware of quite a lot of the work of Wheeler and ADM, and various arguments persuaded us that the geometry of three-dimensional space might well be Riemannian, possess curvature and evolve in accordance with Machian principles. We wanted to find a Machian geometrodynamics, which we did not think would be general relativity. The first task was to select the basic elements of such a theory. What structures should represent instants of time and be the points of the theory's Platonia?

This question was easily answered. Any class of objects that differ intrinsically but are all constructed according to the same rule can form a Platonia. So far, we have considered relative configurations of particles in Euclidean space. There is nothing to stop us considering three-dimensional Riemannian spaces, especially if they are finite because they close up on themselves. This is difficult for a non-mathematician to grasp, but the corresponding things in two dimensions are simply closed, curved surfaces like the surface of the Earth or an egg. The points of Platonia for this case are worth describing. The surface of any perfect sphere is one point; each sphere with a different radius is a different point. Now imagine deforming a sphere by creating puckers on its surface. This can be done in infinitely many ways. There can be all sorts of 'hills' and 'valleys' on the surface of a sphere, just as there are on the Earth and the Moon. And there is no reason why the surface should remain more or less spherical: it can be distorted into innumerable different shapes to resemble an egg, a sausage or a dumbbell. On all of these there can be hills and valleys. Each different shape is just one point in Platonia, and could be a model instant of time. In this case you can form a very concrete image of what each point in Platonia looks like. These are things you could pick up and handle. Note that only the geometrical relationships within the surface count. Surfaces that can be bent into each other without stretching, like the sheet of paper rolled into a tube, count as the same. However, this is a mere technicality. The important thing is that the points of any Platonia are real structured things, all different from one another.

Imagining the points that constitute this Platonia is easy enough. It is much harder to form a picture of Platonia itself because it is so vast and has infinitely many dimensions. Triangle Land has three dimensions, and we can give a picture of it (Figures 3 and 4). But Tetrahedron Land already has six dimensions, and is impossible to visualize. When there are infinitely many dimensions, all attempts at visualization break down, but as mathematical concepts such Platonias do exist and play important roles in both mathematics and physics.

Riemannian spaces are actually empty worlds since they contain nothing that we should recognize as matter. You might wonder in what sense they exist. They certainly exist as mathematical possibilities, and the proof of this was one of the great triumphs of mathematics in the nineteenth century. But they can also contain matter, just like flat familiar Euclidean space. Its properties and existence were originally suggested by the behaviour of matter within it, and evidence for curved space can be deduced through matter as well, as the experimental confirmations of general relativity show. I hope that this disposes of any worries you might have. In fact, the Platonia of three-dimensional Riemannian spaces is well known in the ADM formalism as *superspace* (another Wheeler coining, and not to be confused with a different superspace in superstring theory).

The Platonia that models the actual universe certainly cannot consist of only empty spaces, since we see matter in the world. To get an idea of what is needed, imagine surfaces with marks or 'painted patterns' on them to represent configurations of matter or electric, magnetic or other fields in space. This will hugely increase the number of points in Platonia, since now they can differ in both geometry and the matter distributions. Any two configurations that differ intrinsically in any way count as different possible instants of time and different points of Platonia.

Within classical general relativity, the concept of superspace is not without difficulties, which could undermine my entire programme. Since the issues are decidedly technical, I have put the discussion of them in the Notes. However, I can say here that marrying general relativity and quantum mechanics is certain to require modification of the patterns of thought that have been established in the two separate theories. Superspace certainly arises as a natural concept in the framework of general relativity. The question is whether it is appropriate in all circumstances.

I feel that, when everything has been taken into account, superspace is the appropriate concept, though its precise definition and the kinds of Nows it contains are bound to be very delicate issues. Now, making the assumption they can be sorted out, what can we do with the new model Platonia?

BEST MATCHING IN THE NEW PLATONIA

The key idea in Part 2 is the 'distance' between neighbouring points in Platonia based solely on the intrinsic difference between them. It was obvious to Bruno and me that if we were to make any progress with our more ambitious goal, we should have to find an analogous distance in the new Platonia. We had to look for some form of best matching appropriate in the new arena.

To explain the problem, let me first recall what best matching does and achieves in the Newtonian case of a large (but fixed) number of particles. Each instant of time, each Now, is defined by a relative configuration of them in Euclidean space. We modelled each Now as a 'megamolecule', and compared two such Nows, without reference to any external space or time, by moving one relative to the other until they were brought as close as possible to coincidence as measured by a suitable average. This is where the real physics resides, since the residual difference between the Nows in the best-matching position defines the 'distance' between them in Platonia. Once we possess all such 'distances' between neighbouring Nows, we can determine the geodesics in Platonia that correspond to classical Machian histories. Besides defining these 'distances', the best matching automatically brings the two Nows into the position they have in Newton's absolute space, if we want to represent things in that way.

However, to complete that Newtonian-type picture, we have still to determine 'how far apart in time' the two Nows are. This is the problem of finding the *distinguished simplifier*, the time separation that unfolds the dynamical history in the simplest or most uniform way. As we saw in the final section of Chapter 6, in the discussion of ephemeris time, the choice of distinguished simplifier is unique if we want to construct clocks that will enable their users to keep appointments. Our ability to keep appointments is a wonderful property of the actual world in which we find ourselves, and we must have a proper theoretical understanding of its basis. This is achieved if we insist that a clock is any mechanism that measures, or 'marches in step with', the distinguished simplifier. This is the theory of duration and clocks that Einstein never addressed explicitly. However, the most important thing is that history itself is constructed in a timeless fashion. The distinguished simplifier is introduced after the event to make the final product look more harmonious. Duration is in the eye of the beholder.

In Newtonian best matching, the compared Nows are moved rigidly relative to one another. We could conceive of a more general procedure, but since the Nows are defined by particles in Euclidean space its flatness and uniformity make that an additional complication. We should always try to keep things simple.

However, if we adopt curved three-dimensional spaces, or *3-spaces* as they are often called, as Nows, any best-matching procedure for them will have to use a more general pairing of points between Nows. For example, two 3-spaces (which may or may not contain matter) may have different sizes. It will then obviously be impossible to pair up all points as if they were sitting together in the same space. More generally, the mere fact that both spaces are curved – and curved in different ways – forces us to a much more general and flexible method for achieving best matching.

In a talk, I once illustrated what has to be done by means of two magnificent fungi of the type that grow on trees and become quite solid and firm. For reasons that will become apparent, I called them Tristan and Isolde. Tristan was a bit larger than Isolde, and both were a handsome rich brown, the darkness of which varied over their curved and convoluted surfaces. I wanted to explain how one could determine a 'difference' between the two by analogy with the best-matching for mass configurations in flat space. In some way, this would involve pairing each point on Tristan to a matching point on Isolde. A little reflection shows that the only way to do this is to consider absolutely all possible ways of making the matching.

I took lots of pins, numbered 1, 2, 3, ..., and stuck them in various positions into Tristan. I then took a second set, also numbered 1, 2, 3, ..., and stuck them into Isolde. Since they had similar shapes, I placed the pins in corresponding positions, as best as I could judge. I could then say that, provisionally, pin 1 on Tristan was 'at the same position' as pin 1 on Isolde. All the other points on them were imagined to be paired similarly in a *trial pairing*.

This made it possible to determine a *provisional difference*. For example, I could compare the two fungi using the darkness of their brown surfaces. Alternatively, and much closer to what happens in general relativity, I could compare the curvatures at matching points. The essential point is that some intrinsic property is compared at each pair of matched points, and an average of all the resulting differences is then determined. This average, one number, is the provisional difference. I leave out the mathematical details, which are intricate even though the underlying idea is simple.

This provisional difference is clearly arbitrary since the pairing on which it is based could have been made differently. To find an intrinsic difference that can have real physical meaning, we must now embark – in imagination at least – on an immensely laborious task. Keeping the pins on Tristan fixed, we need to rearrange the pins (reasonably continuously so that the mathematics works) on Isolde in every conceivable way. For each trial pairing of all points on Isolde to all points on Tristan, we must find the provisional difference. We shall know that we have found the best-matching pairing and corresponding intrinsic difference when the provisional difference remains unchanged if we go from the given pairing to any other pairing that differs from it ever so slightly. (In mathematics, the fulfilment of this condition indicates that one has found a maximum, a minimum or a so-called stationary point of the quantity being considered. It turns out that a stationary point is what is found in this case, but that is a mere technicality.) Since there is an immense – indeed infinite – number of ways of changing the pairings, the best-matching requirement imposes a very strong condition. It is impossible to conceive of a more refined and delicate comparison of two things that are different but of the same kind. However, as Bruno and I realized, it is made necessary by the nature of the compared things.

It leads immediately to the *ne plus ultra* of best matching – and rationality.

CATCHING UP WITH EINSTEIN

It was around 1979 that Bruno and I developed the new best-matching idea. We did quite a lot of technical work, and were beginning to get quite hopeful. We knew that we could construct various forms of Machian geometrodynamics, and we began to think that one of them might be a serious rival to general relativity. But it is not easy to beat Einstein, as we were soon to find. This came about through the intervention of another friend, Karel Kuchař, whom I had got to know in 1972, when we had several discussions. Karel is Czech and studied physics at the Charles University in Prague, specializing in relativity. In 1968 he won an award to study at Princeton with John Wheeler, where he quickly established himself as a leading expert in the canonical quantization of gravity (the most straightforward quantization procedure (Box 2) that can be used in the attempt to quantize gravity), in which Dirac and ADM had been the pioneers. Some years later he became a professor of physics at the University of Utah in Salt Lake City, where he still works. Over the years I have profited greatly from discussions with Karel, and certainly would not have been in the position to write this book without assistance from him at some crucial points. However, I hasten to add that Karel is sceptical about my idea that time does not exist at all. As we shall see, general relativity presents a great dilemma. Karel gives more weight to one horn of this dilemma, I to the other.

The issue came into clear focus for me in 1980. In April of that year, Karel gave a memorable review talk at an international conference in Oxford, during which I had an opportunity to discuss with him the ideas that Bruno and I were developing. He invited me to come to Salt Lake City, which I did in the late fall, just in time to see the pale gold of the aspens in the Wasatch mountains. Getting to know Utah and the magnificent deserts of the western United States has been a great bonus from the study of physics for me and my family. But as this is a book about physics, not travel, I had better not digress.

To come straight to the point, it soon became clear in the discussions with Karel that the idea of best matching and the whole way of thinking about duration as a measure of difference were already both contained within the mathematics of general relativity, though not in a transparent form. These facts are still not widely known, mainly, I think, because of a certain inertia. General relativity was discovered as a theory of four-dimensional space-time, and that is still essentially the way it is presented. The fact that it is simultaneously a dynamical theory describing the changes of three-dimensional things is given much less weight. This is why so few people are aware that there is such a deep issue and crisis about the nature of time at the heart of general relativity.

I think that the nature of the problem can be explained to a non-scientist. Here, at least, is my attempt. Figure 29 is a very schematic representation of the three different kinds of four-dimensional space-time that have been considered in this book. As usual, only one of the

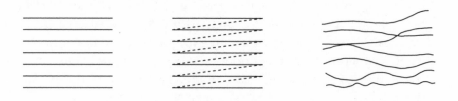

Figure 29 The three different kinds of space-time: on the left, Newtonian space-time, with 'horizontal' Nows; in the middle, Minkowski space-time, with alternative 'tilted' Nows; on the right, the space-time of general relativity, with Nows running in arbitrary directions.

three dimensions of space is shown. It and its material contents are represented by the horizontal direction, while time runs vertically. Thus, the more or less horizontal lines and curves in the three parts of the diagram represent space and its material contents at different 'times'. They are each Nows in my sense. As we have seen, Newtonian space-time is like a pack of ordinary cards. Each card is a Now, and they are all horizontal. I called Minkowski space-time a magical pack of cards because its Nows, or hyperplanes of simultaneity, can be drawn in different ways. Depending on the Lorentz frame that is chosen, different families of parallel Nows are obtained. Time has become relative to the frame. In general relativity, this relativity of time is taken much further: provided the Nows do not cut through the light cone, they can be drawn in an immense number of different ways. It is the complete absence of uniqueness in the way this is done that led Einstein to comment that the concept of Now does not exist in modern physics. However, this reflects the space-time viewpoint. The dynamical viewpoint puts things in a different perspective.

To see this, suppose we consider two neighbouring Nows, as shown in Figure 30, in a space-time that satisfies the equations of general relativity. Each Now is a 3-space with its own intrinsic three-dimensional geometry and material contents embedded within space-time. This four-dimensional space-time has its own geometry too, and permits the construction of 'struts' between the two Nows. The struts are the world lines of bodies that follow geodesics in space-time, leaving the earlier Now along the space-time direction that is perpendicular to it at the

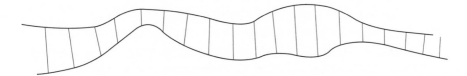

Figure 30 The two continuous curves represent (in one dimension) the two slightly different 3-spaces mentioned in the text; the more or less vertical lines are the 'struts'.

point of departure. Each 'strut' is, so to speak, erected on the first Now. It will pierce the second Now at some point. Taken altogether, such struts uniquely determine a pairing of each point of the first Now with a point of the second Now. They do something else, too. If a clock travels along each strut between its two ends, it will measure the proper time between them as it goes. Because the two Nows have been chosen arbitrarily, the proper time will in general be different for each strut.

What has this to do with best matching? Everything. Imagine mean-minded mathematicians who stick 'pins' like those that I stuck into Tristan and Isolde into the two 3-spaces to identify the two ends of all the struts in Figure 30. The pins carry little flags with the 'lengths' of the corresponding struts – the proper time – along them. However, all this information, which tells us exactly how the two 3-spaces are positioned relative to each other in space-time, is made invisible to other math-ematicians who are 'given' just the two 3-spaces, the Nows with their intrinsic geometries and matter distributions, and set the task of finding the struts' positions and lengths. Will they succeed?

Despite niggling qualifications, the answer is yes. When you unpack the mathematics of Einstein's theory and see how it works from the point of view of geometrodynamics, it appears to have been tailor-made to solve this problem. This was shown in 1962 in a remarkable, but not very widely known paper of just two pages by Ralph Baierlein, David Sharp and John Wheeler (the first two were students of Wheeler at Princeton). I shall refer to these authors, whose paper has the somewhat enigmatic title 'Three-dimensional geometry as a carrier of information about time', as BSW. Initials can become a menace, but the BSW paper is so central to my story that I think they are warranted.

It is the implications of the BSW paper that I discussed with Karel in 1980. They can be quickly summarized. The basic problem that BSW considered was what kind of information, and how much, must be specified if a complete space-time is to be determined uniquely. This is exactly analogous to the question that Poincaré asked in connection with Newtonian dynamics, and then showed that the information in *three* Nows was needed. As we have seen, a theory will be Machian if *two* Nows are sufficient. What BSW showed is that the basic structure of general relativity meets this requirement.

In fact, the all-important Einstein equation that does the work is precisely a statement that a best-matching condition between the two 3-spaces does hold. The pairing of points established by it is exactly the pairing established by the orthogonal struts. In fact, the key geometrical property of space-times that satisfy Einstein's equations reflects an underlying principle of best matching built into the foundations of the theory. I think that Einstein, with his deep conviction that nature is supremely rational, would have been most impressed had he lived to learn about it.

Equally beautiful and interesting is the condition that determines 'how far apart in time' the 3-spaces are. It is closely analogous to the rule by which duration can be introduced as a distinguished simplifier in Machian dynamics and the method by which the astronomers introduced ephemeris time. There is, however, an important difference. In the simple Machian case, the distinguished simplifier creates the same 'time separation' across the whole of space. In Einstein's geometrodynamics, the separation between the 3-spaces varies from point to point, but the principle that determines it is a generalization, now applied locally, of the principle that works in the Newtonian case and explains how people can keep appointments. This is why I say that, quite unbeknown to him, Einstein put a theory of Mach's principle and duration at the heart of his theory.

I go further. The equivalence principle too is very largely explained by best matching. To model the real universe, the 3-spaces must have matter distributions within them. The analogue in two dimensions is markings on bodies or paintings on curved surfaces. When we go through the best-matching procedure, sticking pins into Isolde, it is not only points on her skin that are matched to points on Tristan, but also any tattoos or other decorative markings. All these decorations – matter in the real universe – contribute with the geometry in determining the best-matching position and the distinguished simplifier that holds the 3-spaces apart and creates proper time between them. When this idea is combined with the relativity requirement, the equivalence principle comes out more or less automatically.

Since the equivalence principle is essentially the condition that the law of inertia holds in small regions of space-time, and all clocks rely in one way or another on inertia, this is the ultimate explanation of why it is relatively easy (nowadays at least) to build clocks that all march in step. They all tick to the ephemeris time created by the universe through the best matching that fits it together.

A SUMMARY AND THE DILEMMA

We have reached a crucial stage, and a summary is called for. In all three forms of classical physics – in Newtonian theory, and in the special and general theories of relativity – the most basic concept is a framework of space and time. The objects in the world stand lower in the hierarchy of being than the framework in which they move. We have been exploring Leibniz's idea that only things exist and that the supposed framework of space and time is a derived concept, a construction from the things.

If it is to succeed, the only possible candidates for the fundamental 'things' from which the framework is to be constructed are configurations of the universe: Nows or 'instants of time'. They can exist in their own right: we do not have to presuppose a framework in which they are embedded. In this view, the true arena of the world is timeless and frameless – it is the collection of all possible Nows. Dynamics has been interpreted as a rule that creates histories, four-dimensional structures built up from the three-dimensional Nows. The acid test for the timeless alternative is the number of Nows needed in the exercise. If two suffice, perfect Laplacian determinism holds sway in the classical world. It will have a fully rational basis. There will be a reason for everything, found by examination and comparison of any two neighbouring Nows that are realized. There is perfection in such dynamics: every last piece of structure in either Now plays its part and contributes, but nothing more is needed.

In non-relativistic dynamics, Newton's seemingly incontrovertible evidence for a primordial framework and the secondary status of things can be explained if the universe is Machian. Then the roles will be reversed, things will come first, and the local framework defined by inertial motion will be explained. However, without access to the complete universe such a theory cannot be properly tested. In any case, the Newtonian picture is now obsolete even if it did clarify the issues. In general relativity the situation is much more favourable and impressive, since the best matching is infinitely refined and its effects permeate the entire universe. We can test for them locally. Finding that they are satisfied at some point in space-time is like finding a visiting card: 'Ernst Mach was here'. The strong evidence that Einstein's equations do hold suggests that physics is indeed timeless and frameless.

For all that, the manner in which space-time holds together as a four-dimensional construct is most striking. It is highlighted by the fact that there is no sense in which the Nows follow one another in a unique sequence. This is what, in the Newtonian case, gives rise to the beautifully simple image of history as a curve in Platonia. But in special relativity and, much more strikingly, in general relativity such a unique curve of history is lost. One and the same space-time can be represented by many different curves in Platonia. Even though no extra structure beyond what already exists in Platonia is needed to construct space-time, the way it holds together convinces most physicists that space-time (with the matter it contains) is the only thing that should be regarded as truly existing. They are very loath to accord fundamental status to 3-spaces in the way the dynamical approaches of Dirac, ADM and BSW require. Even though most of them grant that quantum theory will almost certainly modify drastically the notion of space-time, they are still very anxious to maintain the spirit of Minkowski's great 1908 lecture. They are convinced that space and time hang together, and they want to preserve that unity at all costs. Within the purely classical theory, it seems to me that the argument is finely balanced. Perhaps an unconventional image of space-time will show how delicate this issue – space-time as against dynamics – is.

Wagner's opera *Tristan und Isolde* is widely regarded as a highpoint of the Romantic movement in music. General relativity is the *ne plus ultra* of dynamics. More explicitly, the way in which two 3-spaces are fitted together in its dynamical core is like two lovers seeking the closest

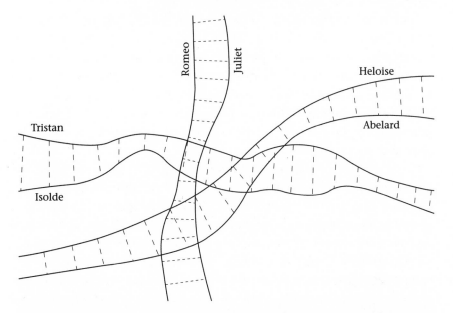

Figure 31 Space-time as a tapestry of interwoven lovers. Given just the 'intrinsic structure' of Tristan and Isolde, the BSW formalism determines in principle all the points on Tristan that will be paired with points on Isolde. The lengths of the struts (proper time between matched points) are obtained as a by-product of the basic problem – finding the 'best position' for the closest possible embrace. They are therefore shown as dashes. The lengths of the struts are local analogues of ephemeris time and, as they separate Tristan and Isolde, are simply the most transparent way of depicting the intrinsic difference between the two of them. The struts between the other pairs of lovers are determined similarly. We can see how the difference that keeps Tristan apart from Isolde is actually part of the body of Romeo (and Juliet). The struts between Romeo and Juliet are drawn with short dashes because they have a space-like separation. Einstein's equations and the best-matching principle hold, however space-time is sliced.

possible embrace. This is the level of refinement at work in the principles that create the fabric of space-time. It is vastly more than just a four-dimensional block. Everywhere we look, it tells the same great story but in countless variations, all interwoven in a higher-dimensional tapestry. This is what Einstein made out of Minkowski's magical pack of cards. Look at space-time one way, and we see Tristan and Isolde hanging, Chagall-like, in the sky. Look another way, and we see Romeo and Juliet, yet another way and it is Heloise and Abelard. All these pairs, each perfect in themselves, are all made out of each other. They and their stories stream through each other. They create a criss-cross fabric of space-time (Figure 31).

It stretches to the limit the notion of substance. For the body of space-time, its fattening in time, is just the way we choose to hold things apart so that the story unfolds simply. At least, it is in Newtonian space-time. All the dynamics – what actually happens – is in the horizontal placing. We pull the cards apart in a vertical direction that we call time as a device for achieving simplicity of representation. Time is the distinguished simplifier. The substance is in the cards. They are the things; the rest is in our mind.

General relativity adds an amazing twist to this seemingly definitive theory of time. Considered alone, Tristan and Isolde are substance, and the separation between them is just the measure of their difference. They cannot come together completely simply because they are different. This difference we call time. But what is representation of difference between Wagner's lovers is part of the very substance of Shakespeare's lovers. Romeo and Juliet would not be what they are if Tristan and Isolde were not held apart by their difference. The time that holds Tristan apart from Isolde is the body of Romeo. This interstreaming of essence and differ-ence all in one space-time is even more remarkable than Minkowski's diagram containing two rods each shorter than the other.

Several profound ideas are unified and taken to the extreme in Figure 31: Einstein's relativity of simultaneity, Minkowski's fusion of time with space, Poincaré's idea that the relativity principle should be realized through perfect Laplacian determinism, Poincaré's idea that duration is defined so as to make the laws of nature take the simplest form possible, and the astronomers' realization that it is measured by an average of everything that changes. Since best matching in general relativity holds throughout the universe in all conceivable directions, both time and space appear as the distillation of all differences everywhere in the uni-verse. Machian relationships are manifestly part of the deep structure of general relativity. But are they the essential part?

If the world were purely classical, I think we would have to say no, and that the unity Minkowski proclaimed so confidently is the deepest truth of space-time. The 3-spaces out of which it can be built up in so many different ways are knitted together by extraordinarily taut inter-woven bonds. This is where the deep dilemma lies. Four decades of research by some of the best minds in the world have failed to resolve it. On the one hand, dynamics presupposes – at the foundation of things – three-dimensional entities. Knowing nothing about general relativity, someone like Poincaré could easily have outlined a form of dynamics that was maximally predictive, flexible, refined and made no use of eternal space or time. Such dynamics, constrained only by the idea that there are distinct things, must have a certain general form. A whole family of theories can be created in the same Machian mould.

On the other hand, a truly inspired genius might just have hit on one further condition. Let dynamics do all those things with whatever three-dimensional entities it may care to start from. But let there be one supreme overarching principle, an even deeper unity. All the three-dimensional things are to be, simultaneously with all their dynamical properties, mere aspects of a higher four-dimensional unity and symmetry.

If certain simplicity conditions are imposed, only one theory out of the general family meets this condition. It is general relativity. It is this deeper unity that creates the criss-cross fabric of space-time and the great dilemma in the creation of quantum gravity. As we shall see, quantum mechanics needs to deal with three-dimensional things. The dynamical structure of general relativity suggests – and sufficiently strongly for Dirac to have made his 'counter-revolutionary' remark – that this may be possible. Yet general relativity sends ambivalent signals. Its dynamical structure says 'Pull me apart', but the four-dimensional symmetry revealed by Minkowski says 'Leave me intact.' Only a mighty supervening force can shatter space-time.

Note added for this printing. New work summarized on p. 358 could significantly change the situation discussed in this final section of the chapter. It suggests that the timeless Machian approach is capable of leading to a *complete* derivation of general relativity and that it is not necessary to presuppose 'a higher tour-dimensional unity and symmetry.' Since this new work has only just been published and has not yet been exposed to critical examination, I decided to leave the original text intact. However, as already indicated in the note at the end of the Preface, this new work does have the potential to strengthen considerably the arguments for the nonexistance of time.

PART 4

Quantum Mechanics and Quantum Cosmology

If the difference between Newtonian and Einsteinian physics is great, quantum mechanics seems separated from both by a chasm. Most accounts of it, however, do not question the framework, essentially absolute space and time, in which it was formulated. They describe how very small systems – mostly atoms and molecules – behave in an external framework. This may make quantum mechanics appear more baffling than need be.

If quantum mechanics is universally true and applies not only to atoms and molecules but also to apples, the Moon, the stars and ultimately the universe, then we ought to consider *quantum cosmology*. What does the quantum mechanics of the universe look like? It cannot be formulated in an external framework. Like classical physics, quantum cosmology needs a description without a framework. We shall see that many apparent differences between classical and quantum mechanics then appear in a different light. What remains is one huge difference. We shall soon begin to get to grips with it.

CHAPTER 12

The Discovery of Quantum Mechanics

About a hundred years ago, a dualistic picture of the world took shape. The electron had just been discovered, and it was believed that two quite different kinds of thing existed: charged particles and the electromagnetic field. Particles were pictured as little billiard balls, possessing always definite positions and velocities, whereas electromagnetic fields permeated space and behaved like waves. Waves interfere, and recognition of this had led Thomas Young to the wave theory of light (Figure 22).

By the end of the nineteenth century, the evidence for the wave theory of light was very strong. However, it was precisely the failure of light, as electromagnetic radiation, to behave in all respects in a continuous wave-like manner that led first Max Planck in 1900 and then Einstein in 1905 to the revolutionary proposals that eventually spawned quantum mechanics. A problem had arisen in the theory of ovens, in which radiation is in thermal equilibrium with the oven walls at some temperature. Boltzmann's statistical methods, which had worked so well for gases, suggested that this could not happen, and that to heat an oven an infinite amount of energy would be needed. The point is that radiation can have any wavelength, so radiation with infinitely many different wavelengths should be present in the oven. At the same time, the statistical arguments suggested that, on average, the same finite amount of energy should be associated with the radiation when in equilibrium. Therefore there would be an infinite amount of energy in the oven – clearly an impossibility. Baking ovens broke the laws of physics! Planck was driven to assume that energy is transferred between the oven walls and the radiation not continuously but in 'lumps', or 'quanta'.

Accordingly, he introduced a new constant of nature, the *quantum of action*, now called *Planck's constant*, because the same kind of quantity appears in the principle of least action. Until Planck's work, it had been universally assumed that all physical quantities vary continuously. But in the quantum world, action is always 'quantized': any action ever measured has one of the values 0, ½*h*, *h*, ³⁄₂*h*, 2*h*, ... Here *h* is Planck's constant. (The fact that half-integer values of *h*, i.e. ½*h*, ³⁄₂*h*, ..., can occur in nature was established long after Planck's original discovery. By then it was too late to take half the original quantity as the basic unit.) The value of *h* is tiny.

Most people are familiar with the speed of light, which goes seven times round the world in a second or to the Moon and back in two and a half seconds. The smallness of Planck's constant is less well known. Comparison with the number of atoms in a pea brings it home. Angular momentum is an action and can be increased only in 'jerks' that are multiples of *h*. Suppose we thread a pea on a string 30 cm long and swing it in a circle once a second. Then the pea's action is about 10^{32} times *h*. As we saw, the atoms in a pea, represented as dots a millimetre apart, would comfortably cover the British Isles to a depth of a kilometre. The number 10^{32}, represented in the same way, would fill the Earth – not once but a hundred times. Double the speed of rotation, and you will have put the same number of action quanta into the pea's angular momentum. It is hardly surprising that you do not notice the individual 'jerks' of the *h*s as they are added.

When people explain how our normal experiences give no inkling of relativity and quantum mechanics, the great speed of light and the tiny action quantum are often invoked. Relativity was discovered so late because all normal speeds are so small compared with light's. Similarly, quantum mechanics was not discovered earlier because all normal actions are huge compared with *h*. This is true, but in a sense it is also misleading. For physicists at least, relativity is completely comprehensible. The mismatch between the relativistic world and its non-relativistic appearance to us is entirely explained by the speed of light. In contrast, the mere smallness of Planck's constant does not fully explain the classical appearance of the quantum world. There is a mystery. It is, I believe, intimately tied up with the nature of time. But we must first learn more about the quantum.

Einstein went further than Planck in embracing discreteness. His 1905 paper, written several months before the relativity paper, is extraordinarily prescient and a wonderful demonstration of his ability to draw far-reaching conclusions from general principles. He showed that in some respects radiation behaved as if it consisted of particles. In a bold move, he then suggested that 'the energy of a beam of light emanating from a certain point is not distributed continuously in an ever increasing volume but is made up of a finite number of indivisible quanta of energy that are absorbed or emitted only as wholes'. Einstein called the putative particles *light quanta* (much later they were called *photons*). In a particularly beautiful argument, Einstein showed that their energy E must be the radiation frequency ω times Planck's constant: $E = \hbar\omega$. This has become one of the most fundamental equations in physics, just as significant as the famous $E = mc^2$.

The idea of light quanta was very daring, since a great many phenomena, above all the diffraction, refraction, reflection and dispersion of light, had all been perfectly explained during the nineteenth century in terms of the wave hypothesis and associated interference effects. However, Einstein pointed out that the intensity distributions measured in optical experiments were invariably averages accumulated over finite times and could therefore be the outcome of innumerable 'hits' of individual light quanta. Then Maxwell's theory would correctly describe only the averaged distributions, not the behaviour of the individual quanta. Einstein showed that other phenomena not belonging to the classical successes of the wave theory could be explained better by the quantum idea. He explained and predicted effects in ovens, the generation of cathode rays by ultraviolet radiation (the photoelectric effect), and photoluminescence, all of which defied classical explanation. It was for his quantum paper, not relativity, that Einstein was awarded the 1921 Nobel Prize for Physics.

The great mystery was how light could consist of particles yet exhibit wave behaviour. It was clear to Einstein that there must be some statistical connection between the positions of the conjectured light quanta and the continuous intensities of Maxwell's theory. Perhaps it could arise through significantly more complicated classical wave equations that described particles as stable, concentrated 'knots' of field intensity. Maxwell's equations would then be only approximate manifestations of this deeper theory. Throughout his life, Einstein hankered after an explanation of quantum effects through classical fields defined in a space-time framework. In this respect he was surprisingly conservative, and he famously rejected the much simpler statistical interpretation provided for his discoveries by the creation of quantum mechanics in the 1920s.

In the following years, Einstein published several important quantum papers, laying the foundations of a quantum theory of the specific heats of solids. However, the next major advance came in 1913 with Danish physicist Niels Bohr's atomic model. It had long been known that atoms emit radiation only at certain frequencies, called *lines* because of their appearance in spectra. These spectral lines, which had been arranged purely empirically in regular series, were a great mystery. Everyone assumed that each line must be generated by an oscillatory process of the same frequency in the atoms, but no satisfactory model could be constructed.

Bohr found a quite different explanation. In a famous experiment, the New Zealander Ernest Rutherford had recently shown that the positive charge in atoms (balanced by the negative charge of the electrons) was concentrated in a tiny nucleus. This discovery was itself very surprising and is illustrated by a well-known analogy. If the space of an atom – the region in which the electrons move – is imagined as being the size of a cathedral, the nucleus is the size of a flea. Bohr supposed that an atom was something like the solar system, with the nucleus the 'Sun' and the electrons 'planets'.

However, he made a seemingly outrageous ad hoc assumption. Using the electrostatic force for the known charges of the electron and positive nucleus, he calculated the electron orbits in Newtonian mechanics for the hydrogen atom, which has only one electron. Each such orbit has a definite angular momentum. Bohr suggested that only orbits for which this angular momentum is some exact multiple of Planck's constant, i.e. 0, h, $2h$, ..., can occur in nature. These orbits also have definite energies, now called *energy levels*. He made the further equally outrageous conjecture that radiation in spectral lines arises when an electron 'jumps' (for some unexplained reason) from an orbit with higher energy to one with lower energy. He suggested that the difference E of these energies is converted into radiation with frequency ω, determined by the relation $E = h\omega$ found by Einstein for the 'lump of energy' associated with radiation of frequency ω. Thus, according to Bohr's theory, an atom emits a light quantum (photon) of a well-defined energy by jumping from one orbit to another.

For hydrogen atoms, it was easy to calculate the energy levels and hence the frequencies of their radiation. Subject to certain further conditions, Bohr's theory had an immediate success. His hotchpotch of

Newtonian theory and strange quantum elements had hardly explained the enigmatic spectral lines, but it did predict their frequencies extraordinarily well, and there could be no doubting that he had found at least some part of a great truth.

During the next decade the Bohr model was applied to more and more atoms, often but not always with success. It was clearly ad hoc. The need for an entirely new theory of atomic and optical phenomena based on consistent quantum principles became ever more transparent, and was keenly felt. Finally, in 1925/6 a complete quantum mechanics was formulated – by Werner Heisenberg in 1925 and Erwin Schrödinger in 1926 (and called, respectively, *matrix mechanics* and *wave mechanics*). At first, it seemed that they had discovered two entirely different schemes that miraculously gave the same results, but quite soon Schrödinger established their equivalence.

Heisenberg's scheme, or *picture*, is based on abstract algebra and is often regarded as giving a truer picture. In the form in which quantum theory currently exists, it is more flexible and general. Unfortunately, it is rooted in abstract algebra, making it very difficult to describe in intuitive terms. I shall therefore use the Schrödinger picture. Luckily, this will not detract from what I want to say. In fact, one of the main ideas I want to develop is that the Schrödinger picture is actually more fundamental than the Heisenberg picture, and is the only one that can be used to describe the universe quantum-mechanically. Many physicists will be sceptical about this, but perhaps this is because they study phenomena in an environment and do not consider how local physics might arise from the behaviour of the universe as a whole.

Schrödinger's work developed out of yet another revolutionary idea, put forward by the Frenchman Louis de Broglie in 1924. It finally overthrew the dualistic picture of particles and fields that had crystallized at the end of the nineteenth century. Einstein had already shown that the electromagnetic field possessed not only wave but also particle attributes. De Broglie wondered whether, since light can behave both as wave and particle, *might not electrons do the same*? Together with its position, the most fundamental property of a particle of mass m and velocity v is its momentum, mv. De Broglie assumed that particles are invariably associated with waves with wavelength λ related by Planck's constant to their momentum: $\lambda = h/mv$.

He applied this idea to Bohr's model. At each energy level, the electron has a definite momentum and hence a wavelength. We can imagine moving round an orbit, watching the wave oscillations. In general, if we start from a wave crest, the wave will not have returned to a crest after one circuit. De Broglie showed that crest-to-crest matching, or *resonance*, would happen only for the orbits with quantized angular momentum that figured so prominently in the Bohr model.

Although he had not, strictly, made any new discovery, his proposal was suggestive. It restored a semblance of unity to the world – both electrons and the electromagnetic field exhibited wave and particle properties. De Broglie's thesis was sent to Einstein, who was impressed and drew attention to its promise. Schrödinger got the hint, and, as they say, the rest is history. During the winter of 1925/6 and the following months he created wave mechanics. This will be the subject of the following chapters.

In 1927 de Broglie's conjecture was brilliantly confirmed for electrons first in an experiment by the Englishman George Thomson, and then in a particularly famous experiment by the Americans Clinton Davisson and Lester Germer. These experiments paralleled those made about a decade and a half earlier by the German physicist Max von Laue, in which he had directed X-rays onto crystals and observed very characteristic diffraction patterns, from which the structure of the crystals could be deduced. The patterns were explained in terms of the interaction of waves with the regular lattice of the atoms forming the crystals. They demonstrated graphically the wave-like behaviour of the electromagnetic field (X-rays are, of course, electromagnetic waves, like light, but with much higher frequency and shorter wavelength). In the 1927 experiments, electrons were directed onto crystals, and diffraction patterns identical in nature to those produced by X-rays were seen. Thus, the particle nature of electrons was observed long before their wave nature was suspected. With light it was the other way round – wave interference was observed a century before Einstein suspected that light could have a particle aspect too.

Although it was now clear that both light and electrons exhibited wave–particle duality, there were important differences between them. A brief description of the picture as it now appears will help. All particles are associated with fields, and can be described as excitations of those fields. To get some idea of what this means, we can liken the particles to water waves, which are excitations of undisturbed water. However, the analogy is only partial. The classic example of a particle associated with a wave is the photon, which is an excitation of the Maxwell field. Fields and associated particles of different kinds exist. There are fields described by a single number at each point, called scalar fields, and vector fields, which are described by three numbers. Scalar fields represent a simple intensity, while the vector fields – such as Maxwell's field – are a kind of 'directed' intensity. In general relativity we also encountered tensors. Mathematically, scalar, vector and tensor fields belong to one family and obey the same kind of rule under rotations of the coordinate system. In particular, after one rotation they return to the values they had before. However, in 1927 yet another sensational quantum discovery was made, this one by Dirac. He found a quite different family of fields, called *spinor fields*, which are associated with electrons and protons (as well as many other particles). In their case, one rotation of the coordinate system brings them back to *minus* the value they had before, and two rotations are needed to restore their original value. Dirac found spinors by trying to make the newly discovered quantum principles compatible with relativity, and achieved a spectacular success even though it was subsequently found that his arguments were not totally compelling. However, the main point is that electrons are associated with a spinor field, photons with a vector field.

Both electrons and photons can, depending on the circumstances, exhibit wave or particle behaviour. Otherwise they behave very differently. Many photons can be present simultaneously in the same state (a state being a characteristic set of properties of particles, such as position and direction of motion), but for electrons this is impossible – there can be at most one in any given state. The two kinds of particle have different statistical behaviour, so-called Fermi–Dirac statistics for electrons and Bose–Einstein statistics for photons. In fact, there are now known to be many different particles, each with an associated field. They satisfy either Fermi–Dirac statistics, and are thus called *fermions*, or Bose–Einstein statistics, in which case they are called *bosons*. In addition, nearly all particles have an antiparticle. An antiparticle is identical to the original particle in some respects, but opposite to it in others; in particular, a particle and its antiparticle always have opposite charges.

In many ways, the story of fundamental physics during the last seventy years has been the discovery of particles and the understanding of the manner in which they interact. All particles that have so far been discovered – there is a whole 'zoo' of them – are either spinor or vector particles.

Ironically, particles corresponding to the simplest scalar fields have not yet been discovered, though it is confidently believed that they will be soon, mainly on the grounds of indirect but rather persuasive theoretical arguments. Currently, an immense amount of work is being done in the attempt to unify the two broad categories of particles – fermions and bosons – by means of an idea called *supersymmetry*. In the last two or three years, there has been another great surge of excitement in the field of superstring theory. This combines the idea of supersymmetry with the idea that the complete 'zoo' of particles known at present are simply different manifestations of the vibrations of a string, much as a violin string can vibrate at its different harmonics. This is the dream of the *theory of everything* (TOE). Some readers may be familiar with these ideas, originally embodied in the acronym GUT – grand unified theory. This was the aim of physicists who wished to describe within a single, unified theoretical framework all the forces of nature except gravity (long recognized as especially difficult to include). More recent, and more ambitious since it aims to include gravity, is the quest for the big TOE.

I am not going to make any attempt to discuss this work, nor will I try to explain the connection between a particle and its associated field. If a theory of everything is found, it may well change the framework of physics. We may find ourselves in a quite new arena and have to change our ideas about space and time yet again. However, as of now I believe we can glimpse the outlines of an arena large enough to accommodate not only the present 'zoo' but also whatever entities some putative theory of everything will come up with. The arena I have in mind is vast and timeless. I see it not as a rival to the theory of everything, but as a general framework in which such a theory can be formulated.

Now it is time to talk about the ideas that Schrödinger introduced in the winter of 1925/6. That was when the door was opened onto the vast arena.

CHAPTER 13

The Lesser Mysteries

INTRODUCTION

Most accounts of quantum mechanics concentrate on the simplest situations – the behaviour of a single particle. That is already very surprising. But the really mysterious properties come to light only in composite systems of several particles, whose behaviour can become bafflingly correlated. The situation is currently very exciting because experimentalists are now able to study two widely separated but strongly correlated particles. Their observations confirm quantum mechanics brilliantly but stretch human intuition to the limit. How can such things happen in space and time? And what unbelievable scenarios will a quantum universe present?

I suspect that the present astonishment exists because most quantum theoreticians do not think enough about quantum cosmology. The first issue is its arena. Quantum mechanics is currently presented in a hybrid framework of two arenas at once. One is an abstract mathematical construct known as Hilbert space, but its elements are essentially defined by absolute space and time, which comprise the second arena. Quantum mechanics takes both for granted. But they provide only a dubious foundation for quantum cosmology. Clarity cannot be achieved until this hybrid state is ended: the space-time framework must go. The answer to the question of how such things can happen in space and time is that they do not. They neither happen nor are they to be found in space and time. But these things are, and their being is in Platonia, which must replace the Hilbert space erected on the shaky foundations of absolute space and time. That, at least, is my view.

My account of wave mechanics will aim to show that the demise of space and time is inevitable. We shall first see how a single particle is described in space and time, and then see what happens when we try to describe the universe. Space and time 'evaporate', and we are left with the one true arena – timeless Platonia. In this arena, quantum mechanics seems to me to take on a totally transparent form. Whether we can believe in it is another matter.

THE WAVE FUNCTION

Every account of quantum mechanics includes the famous two-slit experiment, and mine is no exception (Box 11). Differences come later. The two-slit experiment is to quantum mechanics what the Michelson–Morley experiment is to relativity. The facts are simple, and show that a radical change is unavoidable. The great beauty is that the bare experimental facts directly suggest the need for and the basic form of wave mechanics.

BOX 11   The Two-Slit Experiment

If a beam of photons or electrons, all with the same energy, encounters a slit in a barrier and then impinges on a screen behind it, individual localized 'hits' invariably occur (Figure 32). This is so even if the beam has a very low density, so that at most one particle at a time is passing through the system. This strongly suggests that individual particles leave the beam generator, pass through the slit, and strike the screen. The impacts have a characteristic distribution over a region.

Now introduce a second identical slit in the barrier (Figure 33). The interpretation of the first experiment in terms of individual particles yields an unambiguous prediction for what will happen. The argument is as follows. All particles travel towards the barrier at right angles to it, and can be assumed to be uniformly distributed in space. The pattern behind a single slit is presumably created by the interaction between the particles and the slit as they pass through it. Entering the slit at different positions, the particles will have different deflections and will thus strike the screen at different points. When two slits are open, each should have an effect identical to that of the single slit, so the combined pattern should be simply the sum of the effects of two single slits.

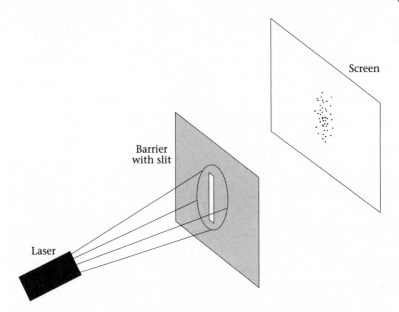

Figure 32 The distribution of hits behind one slit.

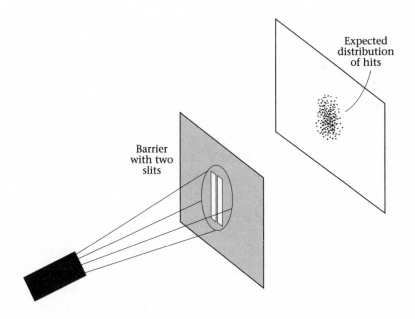

Figure 33 The expected distribution of hits behind two slits.

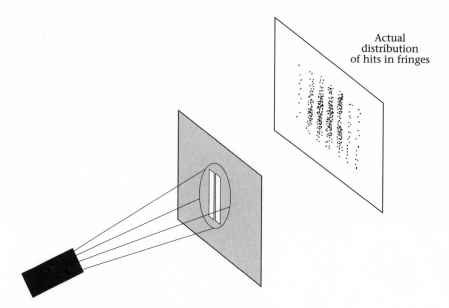

Figure 34 The actual distribution behind two slits.

Nothing remotely resembling this is observed. The hits are distributed in the bands or *fringes* (Figure 34) characteristic of the interference that led Young to the wave theory of light (Figure 22). When, in the nineteenth century, it was believed that these fringes are built up continuously, and not in individual 'hits', it seemed that only a wave field could produce them.

In the absence of a detailed theory, the pattern observed behind a single slit can be explained equally well by particles or waves. But the pattern behind two slits seems totally inexplicable on the assumption of particles. For surely a particle can pass through only one slit, and what it does then will depend solely on the properties of that slit. It cannot 'know' whether the other slit is open or closed and change its behaviour accordingly. Moreover, we can do similar experiments with many slits of different shapes and sizes. Invariably, wave theory correctly predicts the pattern produced on the screen. As far as the total intensity pattern is concerned, there is no way to explain it except by a wave theory.

Yet the patterns are always built up by individual 'hits'. This is extraordinarily strong evidence for particles. But if particles are creating the patterns, they must somehow explore all the slits at once. They must do what the very concept of a particle denies – be everywhere at once. Moreover, this ability to be present at several places at once gives rise to *self-interference*. Dirac put it memorably: 'Each photon ... interferes only with itself.' It is an important observational fact that the possibility for interference to occur continues until something like the screen forces the particle 'to reveal itself'.

As long as the particle is not forced to make a choice, its behaviour in quantum mechanics is described by what Schrödinger called a *wave function*, which he denoted by the Greek letter psi, ψ, and this has become traditional. Sometimes the capital is used: Ψ. I shall use this suitably grander capital in quantum cosmology, keeping ψ for the things that happen in laboratories. The wave function is like an intensity. If x is a point in space, $\psi(x)$ is the value of ψ at x. In general ψ has a different value for each x. The wave function represents something completely new in physics. A further novelty is that the wave function is not an ordinary number, as it would be for a simple intensity, but a *complex number* (Non-mathematicians should not get alarmed: it will be quite sufficient to think of a complex number as a pair of ordinary numbers. 'Complex' in this context means 'composite', not 'complicated'.)

The status of the wave function is contentious to say the least. Some claim it merely represents knowledge, while others want to make it as physical as Faraday's magnetic field. As I see things, the wave function is incorporeal (not some physical thing like a field or particle) and establishes a *ranking* of things. The real things are the points of Platonia, the instants of time. Quantum cosmology – at least in one embryonic form – will associate a value of Ψ (note the capital) with each point of Platonia. To emphasize how different the wave function is, I like to think of it as some 'mist' that hangs or hovers over Platonia, its intensity varying from point to point.

Actually, there are two mists because the wave function, being complex, contains two numbers, which are its two *components*. I shall call them the *red mist* and *green mist*, respectively. I shall also introduce a third number, calling it the *blue mist*. The intensity of this third mist is determined by the two primary components as the sum of the squares of the red and green intensities. This is the mist mentioned in the early chapters. Those in the know will recognize the three mists as the real and imaginary parts of the wave function and the square of its amplitude.

The prominence that I give to these mists could be regarded by most theoretical physicists (above all Dirac and Heisenberg, were they still alive) as a one-sided, if not to say distorted and naive picture of quantum mechanics. The mists (as opposed to things called operators) are not particularly appropriate for talking about most quantum experiments currently performed in laboratories. However, the experiment I have in mind is not done in a laboratory. It is what the universe does to the instants of time. For this experiment, the one that really counts, I think the language of mists is appropriate. Those who disagree might have second thoughts if they really started to think of how inertial frames and duration arise. I come back to these issues later.

I shall now give, in familiar space-time terms, a quantum-mechanical account of the two-slit experiment (Figure 35). At an initial time, the wave function associated with a particle is in a 'cloud' well to the left of the barrier. Inside the cloud, ψ is not zero. Outside, it is zero. As time passes, this cloud moves to the right and, in general, changes its shape. It *evolves* (in accordance with some definite rules). Typically, it 'spreads'. At the barrier, some of the cloud is reflected back to the left but some passes through the two slits. Initially there are two separate clouds, but they spread rapidly if the slits are narrow, and soon overlap. Characteristic wave interference occurs. Thus, when the merged wave reaches the screen, ψ is not the same everywhere, and fringes can form. In fact, the best fringes are formed by a steady 'stream' of wave function, not a cloud.

INTERPRETING THE WAVE FUNCTION

The question now arises: where will the particle in Figure 35 be observed? The answer, given already by the German physicist Max Born in 1926, is that ψ determines, through the intensity of the blue mist, the *probability* of where the particle will be observed. The blue mist enables you to guess where the particle will 'hit' – twice the intensity means twice the probability.

There are many mysteries in quantum mechanics, and the first is the probabilities. We can send identical clouds through the slits many times. The fringe patterns are always exactly reproduced, but the hits are distributed randomly. Only after many 'runs' does a pattern of hits build up. The blue mist gives that pattern. Where its intensity is high, many hits occur; where it is low, few; where it is zero, none. Quantum mechanics

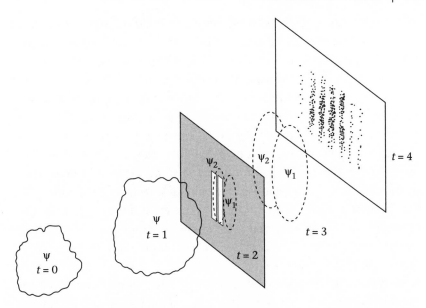

Figure 35 A 'cloud' of wave function ψ approaching two slits (at $t = 0$ and $t = 1$), passing through them, dividing into two (at $t = 2$), spreading and overlapping ($t = 3$) and impinging on a screen ($t = 4$).

determines these probabilities perfectly, but says nothing about where the individual hits will occur.

Einstein found this decidedly disturbing. He could not believe that God reaches for a die every time physicists set up such an experiment and force the particle to show up somewhere. For that is what standard quantum mechanics implies – brute chance determines outcomes. But there are even more puzzling things. It is worth saying that quantum mechanics has a remarkably beautiful and self-contained structure. Examined mathematically, it is a very harmonious whole. It is hard to see how its structure could be modified naturally to make it determine where individual hits occur, especially when relativity is taken into account.

The next mystery is the *collapse of the wave function*. Just before the particle hits the screen, its ψ can be spread out over a large region. What happens to ψ when the particle is suddenly found somewhere? The standard answer is that the wave is instantaneously annihilated everywhere except where the particle is now known to be.

If we want to determine what now happens, we have to start afresh from a small, reduced cloud. The large cloud has been 'collapsed' and has no more relevance. This too provokes much puzzling, especially for those (like Schrödinger in 1926) who would wish to think of ψ as something real, a density of charge, say. How can something real disappear instantaneously? Nothing in the equations describes the collapse – it is simply postulated. Lawful evolution, in accordance with the rules (equations) of quantum mechanics, continues until an observation is made, but then the rules are simply set aside. Quite different rules apply in *measurements*, as they are called. (In quantum mechanics, the term 'measurement' is used a very precise way. It means that some definite arrangement of instruments is used to establish the value of some physical property – say the speed or position of a particle.) The abrupt and schizophrenic change of the rules when measurements are made is a major part of the notorious *measurement problem*. There are rules for evolution and rules for measurement – and they are even more different than chalk and cheese. Nevertheless, both are excellently confirmed, though we have to be careful when saying that the collapse is instantaneous, and even when it occurs.

STATES WITHIN STATES

Just as mysterious as the rule change when measurements are made is a certain mutual exclusivity about the kinds of measurement that can be made. So far, I have talked only about particle positions. However, we can also measure other quantities – for example, a particle's energy, momentum or angular momentum. It is particularly fascinating that information about them all is coded at once in ψ. This is another big difference from classical mechanics.

Imagine a perfect sinusoidal wave that extends with constant wavelength from infinity to infinity. For the moment, suppose that it is 'frozen', like the wave patterns you see in damp sand at low tide. Let me call this the red wave, because it represents the red mist. Now imagine another identical though green wave, shifted forward by a quarter of a wavelength relative to the red wave (Figure 36). Then the red peaks lie exactly at the green wave's nodes, where the green wave has zero intensity. As time passes, the red and green waves move to the right, maintaining always their special relative positioning. A wave function in this special form represents a particle that has a definite momentum: if it

Figure 36 The wave function of a particle with a definite momentum.

hit something, it would transmit a definite impulse to it. A particle with the opposite momentum is represented similarly, but travels in the opposite direction and has the green peaks a quarter of the wavelength behind the red peaks. According to the quantum rules, the particle has a definite momentum because its ψ has a definite wavelength and is perfectly sinusoidal. Such wave functions give the best interference effects in two-slit experiments. They are called momentum *eigenstates*. (The German word *eigen* means 'proper' or 'characteristic'.)

The striking thing about this situation is that the probability for the position of the particle, given by the sum of the squares of the red and green intensities, is completely uniform in space. The reason is that for two sinusoidal waves displaced by a quarter of a wavelength, this sum is always 1 if the wave's amplitude (its height at the peaks) is 1. This is a consequence of the well-known trigonometric relation $\sin^2 A + \cos^2 A = 1$, which itself is just another expression of Pythagoras' theorem. Thus, for a particle in this state, we have absolutely no information about its position, but we do know that it has a definite momentum.

So far we have considered waves of only one wavelength. However, we can add waves of different wavelengths. Whenever waves are added, they interfere, enhancing each other here and cancelling out there. By playing around with waves of different wavelengths we can make a huge variety of patterns (Figure 37 is an example). In fact the French mathematician Joseph Fourier (one of Napoleon's generals) showed that more or less any pattern can be made by adding, or *superposing*, sinusoidal waves appropriately. Any wave pattern created in this way and concentrated in a relatively small 'cloud' is called a *wave packet*. The same pattern can be made by superpositions of quite different kinds. The primary meaning of ψ is that its value at x determines, through the squares of its two intensities, the probability that the particle will be 'found' at x. Now, a 'cloud'

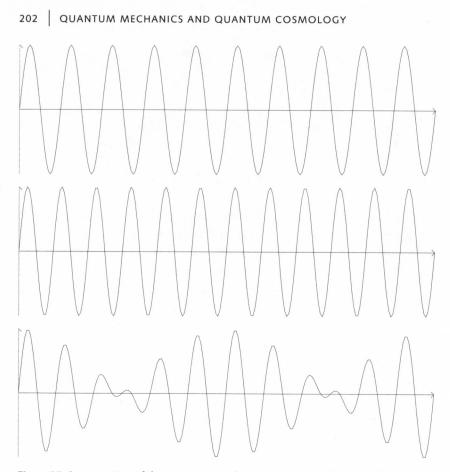

Figure 37 Superposition of the two waves at the top gives rise to the very different wave pattern at the bottom.

could be so narrow that it becomes a 'spike' at some value of x. The particle can then be at only one place – at the spike. Such a wave function is called a position eigenstate.

Thus, the same wave pattern can be regarded either as a superposition of plane waves or as a superposition of many such spikes added together with different coefficients (Figure 38). Any wave function is a superposition of either position or momentum eigenstates. There is a duality at the heart of the mathematics. What is remarkable – and constitutes the essential core of quantum mechanics in the standard form it was given by Dirac – is that it perfectly reflects a similar duality found in nature. This is where the measurement problem becomes even more puzzling. We need to consider the 'official line', known as the Copenhagen interpretation

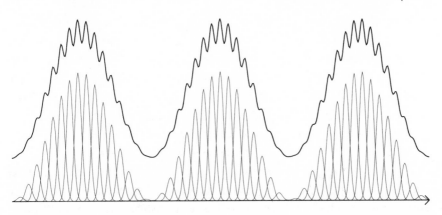

Figure 38 Two 'spiky' wave patterns (thin curves) are superposed to make a much smoother pattern (heavy curve).

because it was established by Heisenberg and Bohr at the latter's institute in Copenhagen shortly after the creation of quantum mechanics.

THE COPENHAGEN INTERPRETATION

The wave function of a particle is assumed to be a maximal representation of its physical state. It codes everything that can ever be deduced about the particle at an instant. Using it, we can predict the outcomes of experiments performed on the particle. There are two cardinal facts about these predictions. First, they are probabilistic. Only if, for example, the particle is in a momentum eigenstate (represented by the two special plane waves described above) will measurement of the momentum confirm that the particle has the corresponding momentum. If it is in a superposition of momentum eigenstates, then any one of the momenta in the superposition may be found as a result of the measurement. The probabilities for them are determined by the strengths with which the corresponding momentum eigenstates are represented in the superposition.

It is a basic Copenhagen tenet that the probabilistic statements reflect a fundamental property of nature, not simply our ignorance. It is not that before the measurement the particle does have a definite momentum and we simply do not know it. Instead, all momenta in the superposition are present as potentialities, and measurement forces one of them to be actualized. This is justified by a simple and persuasive fact. If we do not perform measurement but instead allow ψ to evolve, and only later make some measurement, then the things observed later (like the two-slit fringes) are impossible to explain unless all states were present initially and throughout the subsequent evolution. Outcomes in quantum mechanics are determined by chance at the most fundamental level. This is the scenario of the dice-playing God that so disturbed Einstein.

If anything, the second cardinal fact disturbed him even more. There seems to be a thoroughgoing indefiniteness of nature even more radical than the probabilistic uncertainties. As we have seen, one and the same state can be regarded as a superposition of either momentum or position eigenstates. It is the way this mathematics translates into physics that is startling. The experimentalist has complete freedom to choose what is to be measured: position or momentum. Both are present simultaneously as potentialities in the wave function. The experimentalist merely has to choose between set-ups designed to measure position or momentum. Once the choice is made, outcomes can then be predicted – and one outcome is actualized when the measurement is made. In fact, the indefiniteness is even greater since other quantities, or *observables* as they are called, such as energy and angular momentum, are also present as potentialities in ψ.

Only one experiment can be made – for position or momentum, say, but not both. Every measurement 'collapses' the wave function. After the collapse, the wave function, which could have been used to predict outcomes of alternative measurements, has been changed irrevocably: there is no going back to the experiment we opted not to perform. It is a very singular business. Whatever observable we decide to measure, we get a definite result. But the observable that is made definite depends on our whim. The many people who, like Einstein, believe in a real and definite world find this immensely disconcerting. What is out there in the world seems to depend on mere thoughts that come into our mind. Most commentators believe that this radical indefiniteness – the possibility to actualize either position or momentum but not both – is the most characteristic difference between classical and quantum physics. In classical physics, position and momentum are equally real, and they are also perfectly definite.

The fact that in quantum mechanics one can choose to measure one but not both of two quantities was called *complementarity* by Bohr. Pairs of quantities for which it holds are said to be complementary.

HEINSENBERG'S UNCERTAINTY PRINCIPLE

Heisenberg's famous uncertainty relation gives quantitative expression to complementarity for position and momentum. De Broglie's relation $\lambda = h/mv = h/p$ determines the wavelength of a particle of momentum $p = mv$, where m is its mass and v its velocity. Now, it follows from Fourier's work on the superposing of waves that a wave packet restricted to a small spatial region contains many waves in a broad spread of wavelengths. To narrow down the spatial positions q, it is necessary to broaden the range of momenta p. Conversely, to get a nearly definite p, we must accept a wide range of positions q.

Mathematically, we can in fact construct wave packets in which the positions are restricted to a small range, from q to $q + \Delta q$, and the momenta to a correspondingly small range, from p to $p + \Delta p$. Any attempt to make Δq smaller necessarily makes Δp larger, and vice versa. Heisenberg's great insight – his uncertainty relation – was the physics counterpart of this mathematics. There is always a minimum uncertainty: the product $\Delta q \Delta p$ is always greater than or, at best, equal to Planck's constant h divided by 4π. If you try to pin down the position, the momentum becomes more uncertain, and vice versa. This is the uncertainty relation. Moreover, a wave packet of minimum dimensions will in general spread: the uncertainty in the position will increase. This is what in quantum mechanics is known as the 'spreading of wave packets'.

Since Planck's constant h is so small, an object like a pea or even a grain of sand can effectively have both a definite position and definite momentum, and the spreading of its wave packet takes place extremely slowly. This explains why all the macroscopic objects we see around us can seem to have definite positions. But though the quantum laws allow objects to be localized in space and to have effectively definite velocities, there is no apparent reason in the equations why this should habitually be so. They also allow – encourage, one might even say – a pea's wave packet to be localized in two or more places at once. Nothing forces ψ to 'localize' around a single point. Einstein used to look at the Moon and ask why we do not see two. It is a real problem. Quantum measurements on microscopic systems are actually designed to create situations in which a macroscopic instrument pointer is, according to the equations, in many places at once. Yet we always see it at only one.

THE ENIGMATIC GEM

We shall come back to this mystery, which is one aspect of another: Hilbert space and transformation theory. If you find this section a bit abstract, don't worry; it is helpful at least to mention these things. In quantum mechanics, position and momentum (and other observables) play a role rather like coordinates – 'grid lines' – on a map. Just as in relativity the coordinates on space-time can be 'painted' in different ways, so too in quantum mechanics there are many mathematically equivalent ways of arranging the coordinates. This was one of Dirac's first great insights, and it led to his *transformation theory*.

According to this, the state of a quantum system is some definite but abstract thing in an equally abstract Hilbert space. The one state can, so to speak, be looked at from different points of view. A Cubist painting might give you a flavour of the idea. In relativity, different coordinate systems on space-time correspond to different decompositions into space and time. In quantum mechanics, the different coordinate systems, or *bases*, are equally startling in their physical significance. They determine what will happen if different kinds of measurement, say of position or of momentum, are made on the system by instruments that are external to the system. The state in Hilbert space is an enigmatic gem that presents a different aspect on all the innumerable sides from which it can be examined. As Leibniz would say, it is a city multiplied in perspective. Dirac was entranced, and spoke of the 'darling transformation theory'. He knew he had seen into the structure of things. What he saw was some real but abstract thing not at all amenable to easy visualization. But the multiplication of viewpoints and the mathematical freedom it furnished delighted him.

In *The Principles of Quantum Mechanics*, a veritable bible for quantum mechanicians, Dirac says that in classical physics 'one could form a mental picture in space and time of the whole scheme' but 'It has become increasingly evident that Nature works on a different plan. Her fundamental laws do not govern the world as it appears in our mental picture in any very direct way ...'. I have quoted these words because, with all respect to the greatness of his discoveries and the clarity of his thought, Dirac may have gone too far with his dismissal of simple mental pictures. But what kind of mental pictures are we talking about here? Dirac was reacting against Einstein and Schrödinger, who longed to form mental pictures in space *and* time. Schrödinger, for example, had commented in his second paper on wave mechanics that some people

> had questioned whether the things that happen in the atom could be incorporated in the space-time form of thought at all. Philosophically, I would regard a final decision in this sense as the same as complete capitulation. For we cannot actually change the forms of thought, and what we cannot understand within them cannot be understood at all. There are such things – but I do not think atomic structure is one of them.

This appeal to ineluctable forms of thought, an echo of the eighteenth-century German philosopher Immanuel Kant's belief that space and time are an a priori framework without which we cannot even begin to form a picture of the world, is doubly ironic. Schrödinger was strongly drawn to the holistic notions of eastern mysticism but would not accept them in his own theory, where they seem inescapable. Even more ironically, he himself changed the forms of thought. He created new mental images just as transparent as the space and time to which he and Einstein clung for dear life. That is the topic of the next chapter.

CHAPTER 14

The Greater Mysteries

SCHRÖDINGER'S VAST ARENA

The true heart of quantum mechanics and the way to quantum cosmology is the way in which it describes composite systems – that is, systems consisting of several particles. It is an exciting, indeed extraordinary story, though it is seldom well told. When Schrödinger discovered wave mechanics, he said it could be generalized and 'touches very deeply the true essence [*wahre Wesen*] of the quantum prescriptions'. But it was not just the Bohr quantization prescriptions that came into focus: at stake here are the rules of creation. A bold claim, but one I hope to justify as the book goes on. First, we have to see how Schrödinger opened the door onto a vast new arena.

The central concept of this book is Platonia. It is a relative configuration space. The new arena that Schrödinger introduced is something similar, a *configuration space* (without the 'relative'). The notion is easily explained. Each possible relative arrangement of three particles is a triangle and corresponds to a single point in the three-dimensional Triangle Land. But now imagine the three particles located in absolute space. Besides the triangle they form, which is specified by three numbers (the lengths of its sides), we now have to consider the location of its centre of mass in absolute space, which requires three more numbers, and also its orientation in absolute space, which also requires three more numbers. Location in Triangle Land needs three numbers, in absolute space it needs nine. Just as each triangle corresponds to one point in three-dimensional Triangle Land, the triangle and its location in absolute space correspond to one point in a nine-dimensional configuration space. The tetrahedron formed by four particles corresponds to one point in six-dimensional Tetrahedron Land and one point in the corresponding twelve-dimensional configuration space. For any Platonia corresponding to the relative arrangements of a certain number of particles, the matching configuration space has six extra dimensions. Schrödinger called such a space a Q, and I shall follow his example. Such a Q is a 'hybrid Platonia', since it contains both absolute and relative elements. This hybrid nature is very significant, as will become apparent.

The most important thing about Schrödinger's wave mechanics is that it is formulated not in space and time, but in a suitably chosen Q and time. This is not apparent for a single particle, for which the configuration space is ordinary space. Since most accounts of quantum mechanics consider only the behaviour of a single particle, many people are unaware that the wave function is defined on configuration space. That is where ψ lives. It makes a huge difference.

An illustration using a plastic ball-and-strut model of molecules may help to bring this home. Imagine that you are holding such a model in some definite position in a room, which can represent absolute space. There are three digital displays – I shall call them ψ meters – that show red, green and blue numbers on the wall. These numbers give the intensities of the three 'mists' represented by ψ for the system at the time considered. Suppose you take just one ball, representing one particle of the system, and detach it from the model. Keeping all the other balls fixed, you can move the one ball around and, courtesy of the ψ meters, see how ψ changes. As you move in each direction in space, each ψ value will change. For each point of space you can find the value of ψ. The blue ψ meter will always tell you the positions for which the probability is high or low. Suppose you do this and then return the ball to its original place.

Now move a second ball to a slightly different position, and leave it there. The ψ meters will change to new values. Once again, explore space with the first ball, watching the ψ meters. The values of ψ will be (in general) quite different. The ψ values on the displays embody information. The amount is staggering. For every single position in space to which you move any one of the other balls, you get a complete new set of values in space for the ball chosen as the 'explorer'. And any ball can be the explorer. Each explorer will have its own distinctive three-dimensional patterns of ψ for every conceivable set of positions of the others.

Now, what is a molecule? When Richard Dawkins described the haemoglobin molecule and its six thousand million million million perfect copies in our body, he said that in its intricate thornbush structure there is 'not a twig nor a twist out of place'. That is in a molecule containing perhaps twenty thousand atoms. But molecules are even more remarkable than that. The twig and the twist are averaged structures corresponding to the most probable configuration in which the molecule will be found. In the Schrödinger picture, the molecule is not just one structure but a huge collection of potentially present structures, each with its own probability.

In fact, the complete structure of complicated protein molecules like haemoglobin cannot be understood solely on the basis of wave mechanics. This is because of the way they are put together from amino acid units. But for simpler molecules, which may still contain many particles, you could (in imagination at least) do what I have just described for the ball-and-strut model. Start with one of the model configurations shown in chemistry textbooks, and look at the ψ meters, especially the blue one. It will give a high reading. Around that highly probable structure are other similar structures, all with a high – but not quite so high – blue intensity. Individual units of the structure – simpler forms of Dawkins's 'twigs' – could be moved as a whole, say by twisting them, from the most probable configuration, and the blue intensity would drop. It would also drop if one atom of the few dozen within the twig were moved from the most-favoured position. The molecule is not just the most probable configuration. It is all possible configurations with their ψ values, held in balance by the laws of wave mechanics. The existence and most-favoured shape of molecules can be understood in no other way.

Contrary to the impression given in many books, quantum mechanics is not about particles in space: it is about systems being in configurations – at 'points' in a Q, or 'hybrid Platonia'. That is something quite different from individual probabilities for individual particles being at different points of ordinary space. Each 'point' is a whole configuration – a 'universe'. The arena formed by the 'points' is unimaginably large. And classical physics puts the system at just one point in the arena. The wave function, in contrast, is in principle everywhere.

This is what I mean by saying that Schrödinger opened the door onto a vast new arena. Compared with Schrödinger's vistas, grander than any Wagnerian entrance into Valhalla, the Heisenberg uncertainty relation for a single particle captures little of quantum mechanics. All revolutions in physics pale into insignificance beside Schrödinger's step into the configuration space Q. Not that he did it happily.

CORRELATIONS AND ENTANGLEMENT

It is not possible to observe the extraordinary quantum arena directly. Some people do not believe it exists at all. To a large degree it has been deduced, or surmised, from phenomena observed in systems of a few particles. Getting clear, direct evidence for the quantum behaviour of single particles was difficult. It was long after Dirac made his memorable remark about each photon interfering with itself that the development of sources which release individual particles with long time intervals between releases confirmed the build-up of interference patterns in individual 'hits'. In the last two decades, it has become possible to create in the laboratory pure quantum states of two particles, whose Q therefore has six dimensions. The quantum predictions, all verified, are not easy to explain in many words, let alone a few, and a serious attempt to do so would take me too far from my main story. The simplest possible illustration is given by two particles moving on a single line; each has a one-dimensional Q, and together they have a two-dimensional configuration space (Figure 39).

As for a single particle, the maximally informative description of a quantum system at any instant t is specified by a complex wave function ψ which, in principle, has a different value at each point of the configuration space. As t changes, ψ changes. All information that can be known about the system at t is encoded in ψ at t, and consists of predictions that can be made about it. Many different kinds of prediction can be made, but they are often mutually exclusive. In a very essential way, the predictions refer to the system, not its parts.

Let us start with position predictions. Just as we did for a single particle, we can form from ψ the sum of the squares of its intensities, finding the intensity of the 'blue mist' (Figure 40). This gives the relative probability that the system will be found at the corresponding point in Q if an appropriate measurement is made. The important thing is that a single point in Q corresponds to positions of both particles. Anyone who has not understood this has not understood quantum mechanics. It is this fact, coupled with complementarity, that leads to the most startling quantum phenomena.

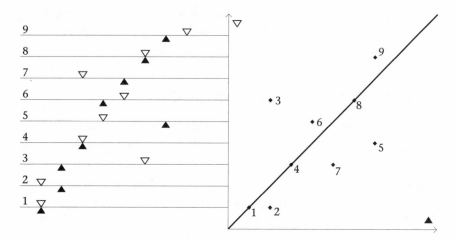

Figure 39 The two-dimensional configuration space Q of two particles on one line. The line is shown in multiple copies on the left. Nine different configurations of the two particles on it are shown. The positions of particles 1 and 2 are indicated by the black and white triangles, respectively. The axes of Q on the right show the distances of particle 1 (horizontal axis) and particle 2 (vertical axis) from the left-hand end of the line. The points on the 45° diagonal in Q correspond to configurations for which the two particles coincide (points 1, 4 and 8). You might like to check how the nine configurations on the left are represented by the nine corresponding points on the right.

In Chapter 3 we imagined tipping triangles out of a bag. That exercise was presented because it mimics one of the ways in which we can interpret a quantum state. Imagine now that the blue mist has the distribution shown in Figure 40. To avoid problems with infinite numbers of configurations, we divide up Q by a grid of cells sufficiently fine that ψ hardly changes within any one of them (Figure 40, on the right). The intensity of the blue mist at the central point of each cell then gives the relative probability of the nearly identical configurations in that cell. On a piece of cardboard, let us depict one of these configurations (as shown on the left in Figures 39 and 40). This will serve as the representative of all the configurations of that cell. For the grid in the figure with 100 cells, there are 100 relative probabilities whose sum should be conveniently large, say a million. Then we shall not distort things seriously by replacing exact relative probabilities like 127.8543... by the rounded-up integer 128.

We now imagine putting into a bag the number of copies of each representative configuration equal to its rounded probability, 128 for example. In quantum mechanics, performing a measurement to deter-

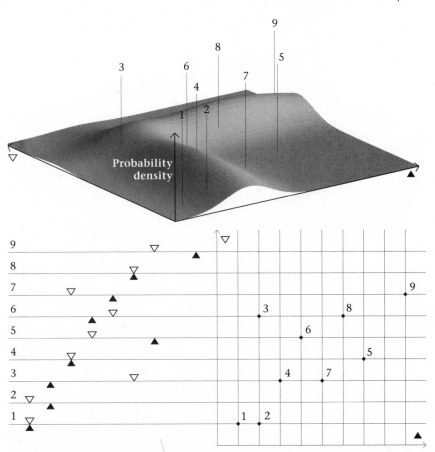

Figure 40 Like Figure 39, this shows nine different configurations of two particles (black and white triangles) on a line and the points corresponding to them in the configuration space Q (on which a grid has been drawn). A possible distribution of the intensity of the blue probability mist is shown as the height of a surface over Q in the top part of the figure (you are seeing the surface in perspective from above, and rotated). In the state of the system shown here, the probabilities for configurations 4, 6 and 9 are high, while 5 has a very low probability.

mine the positions of both particles is like drawing at random one piece of cardboard from the bag. We get some definite configuration. In the process, we destroy the wave function and replace it by one entirely concentrated around the configuration we have found. If we recreate the original wave function, by repeating the operations that we used to set it up, and repeat the experiment millions of times, then the relative frequencies with which the various configurations are 'drawn from the bag' will match, statistically, the calculated relative probabilities.

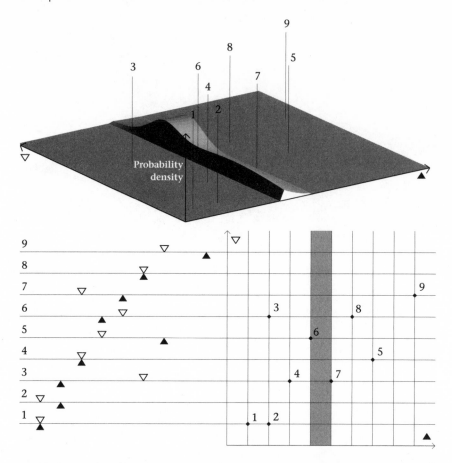

Figure 41 (a) The effect of measuring, for the probability density of Figure 40, the position of the particle represented by the horizontal axis, and finding that it lies in the interval on which the vertical strip stands. All the wave function outside the strip is instantaneously collapsed.

This is only the start. We can select from a menu of different kinds of measurement. For example, we can opt to find the position of only one particle, which has remarkable implications for what we can say about the other one. Suppose first that we measure the position of just one of the particles. According to the quantum rules, this instantaneously collapses the wave function from its original two-dimensional 'cloud' to a one-dimensional profile (Figure 41). The point is that we now know the position of one particle to within some small error, so none of the wave function outside the narrow strip is relevant any longer. It is annihilated.

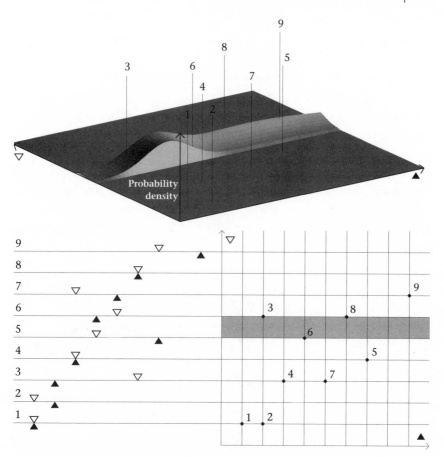

Figure 41 (b) The same for a position measurement of the other particle.

If the particle whose position is measured is represented along the horizontal axis, only a vertical strip of ψ survives (Figure 41(a)); if the position of the other particle is measured, only a horizontal strip survives (Figure 41(b)).

Either profile then gives conditional information. If we know where one particle is, the possible positions of the other are restricted to a narrow strip. The relative probabilities for the position of the second particle are determined by the values of ψ within the strip. Provided we know the original wave function, acquired knowledge about one particle sharpens our knowledge – instantaneously – about the other. This is the place to explain *entangled states*, or *quantum inseparability* (Box 12).

BOX 12   Entangled States

Figure 42 again shows our two-particle Q and two different quantum states. In the *unentangled* state at the bottom, all the horizontal ψ profiles are identical, and so are all the vertical profiles (only their shapes count). Such a wave function is said to be unentangled because if we gain information about particle 1 (black triangles) – that it is at some definite position – we do not gain any new information about particle 2 (white triangles). This is because all the horizontal profiles and all the vertical profiles are identical: they give identical relative probabilities. This is shown by two profiles that result from an exact position measurement of particle 1. Measurement on particle 1 leads to no new information about particle 2.

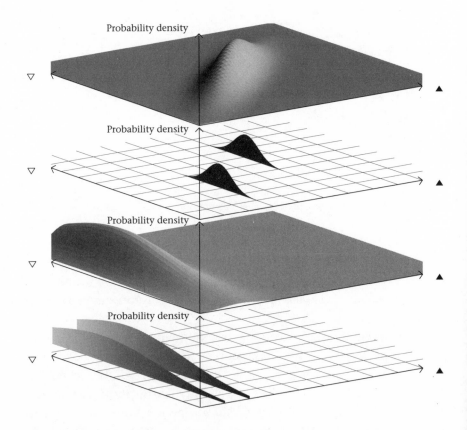

Figure 42 Entangled (top) and unentangled (bottom) states for two particles (black and white triangles).

Much more interesting is the *entangled* state at the top, for which the horizontal and vertical profiles are not identical. Figure 42 shows two profiles that result from exact position measurement of particle 1. They give very different probability distributions for particle 2: the gain in information about particle 2 is considerable. This is typical of quantum mechanics, since virtually all wave functions are entangled to a greater or lesser extent.

A particular feature of entangled states should be noted. Particles normally interact (affect each other) when they are close to each other. For the two-particle Q in Figure 39, the particles coincide on the diagonal line, and the region in which the particles are close to each other and can interact strongly is therefore a narrow strip around that line. However, the wave function of an entangled state may be located completely outside this region – the particles may be very far apart when one of them is observed. Yet the other particle is apparently immediately affected. It can jump to one or other of two hugely different possibilities. Moreover, if the 'ridge' in the top part of Figure 42 is made thinner and thinner, shrinking to a line, then position determination of one particle immediately determines the position of the other to perfect accuracy. Such situations are not easy to engineer for position measurements (and in general will not persist because of wave-packet spreading), but there are analogous situations for momentum and angular-momentum measurements that are easy to set up.

The facts discussed in Box 12 are immensely puzzling if we wish to find a physical and causal mechanism to explain how measurement on one particle can have an immediate effect on a distant particle. As I have already explained, innumerable interference phenomena indicate that, in some sense, the particles are, before any measurement is made, simultaneously present wherever ψ extends. Since there is no restriction on the distance between the particles, any causal effect on the second particle after the first has been observed would have to be transmitted instantaneously. However, relativity theory is supposed to rule out all causal effects that travel faster than the speed of light. Moreover, in the mid-1980s Alain Aspect in Paris performed some very famous experiments in which such wave-function collapses were tested, and the predictions of quantum mechanics confirmed with great accuracy. The experiments were so arranged that any physical effect would have had to be transmitted faster than the speed of light to bring about the collapse.

The situation is actually delicate and intriguing. Relativity absolutely prohibits the transmission of *information* faster than light. But, curiously, wave-function collapse does not transmit information. When information about particle 1 has been obtained by an experimentalist, he or she will know immediately what a distant experimentalist can learn about particle 2. But there is no way such information can be transmitted faster than light. There is no conflict with the rules of relativity, though many physicists are concerned that its 'spirit' is violated.

So far, we have considered only position measurements on a two-particle system. But we can also consider many other measurements, of momentum, for example. Given ψ in Q, we directly obtain predictions for positions. But Dirac's transformation theory enables us to pass to the complementary momentum space, which gives direct predictions for momentum measurements. If the wave function is tightly entangled with respect to momentum, measuring the momentum of particle 1 would immediately tell us the momentum of particle 2. And this despite the fact that before any measurements are made there is considerable uncertainty about the momenta of the particles. However, what is certain is that they are entangled, or correlated. This brings us to the EPR paradox.

THE EPR PARADOX

The nub of the Einstein–Podolsky–Rosen (EPR) paradox, formulated in 1935 by Einstein and collaborators Boris Podolsky and Nathan Rosen, is that two particles can be in a state in which they are perfectly correlated (entangled) as regards both their position and their momentum. The actual example of such a state that EPR found is rather unrealistic, but in 1952 David Bohm, an American theoretical physicist who later worked in London for many years, proposed a much more readily realized state using *spin*, the intrinsic angular momentum associated with quantum particles. Alain Aspect performed his experiments on such a system. What puzzled EPR about their state was that if the position of one particle was measured, the position of the other particle could be immediately established with certainty because of the perfect correlation. Since the second particle, being far away, could not be physically affected by the measurement, but it was known for certain where it would be found, EPR concluded that it must have had this definite property before the measurement on the first particle.

But, it could just as well have been decided to measure momentum. The measurement of one momentum will then instantaneously determine the other momentum with certainty. By the same argument as before, the particle must have possessed that momentum before the measurement on the first particle. Finally, the choice between momentum or position measurement is a matter of our whim, about which the second particle can know nothing. The only conclusion to draw is that the second particle must have possessed definite position *and* momentum before any measurements were made at all. However, according to the fundamental rules of quantum mechanics, as exemplified in the Heisenberg uncertainty principle, a quantum particle cannot possess definite momentum and position simultaneously. EPR concluded there must be something wrong – quantum mechanics must be incomplete.

Niels Bohr actually answered EPR quite easily, though not to everyone's satisfaction. His essential point was that quantum mechanics predicts results made in a definite experimental context. We must not think that the two-particle system exists in its own right, with definite properties and independent of the rest of the world. To make position or momentum measurements, we must set up different instruments in the laboratory. Then the total system, consisting of the quantum system and the measuring system, is different in the two cases. Nature arranges for things to come out differently in the two cases. Nature is holistic: it is not for us to dictate what Nature is or does. Quantum mechanics is merely a set of rules that brings order into our observations. Einstein never found an answer to this extreme operationalism of Bohr, and remained deeply dissatisfied.

I feel sure that Bohr got closer to the truth than Einstein. However, Bohr too adopted a stance that I believe is ultimately untenable. He insisted that it was wrong to attempt to describe the instruments used in quantum experiments within the framework of quantum theory. The classical world of instruments, space and time must be presupposed if we are ever to talk about quantum experiments and communicate meaningfully with one another. Just as Schrödinger made his Kantian appeal to space and time as necessary forms of thought, Bohr made an equally Kantian appeal to macroscopic objects that behave classically. Without them, he argued, scientific discourse would be impossible. He is right in that, but in the final chapters I shall argue that it may be possible to achieve a quantum understanding of macroscopic instruments and their interaction with microscopic systems. Here it will help to consider why Einstein thought the way he did.

Referring to their demonstration that distant measurement on the first system, 'which does not disturb the second in any way', nevertheless seems to affect it drastically, EPR commented that 'No reasonable definition of reality could be expected to permit this.' These words show what is at stake – it is the atomistic picture of reality. Despite the sophistication of all his work, in both relativity and quantum mechanics, Einstein retained a naive atomistic philosophy. There are space and time, and distinct autonomous things moving in them. This is the picture of the world that underlies the EPR analysis. In 1949 Einstein said he believed in a 'world of things existing as real objects'. This is his creed in seven words. But what are 'real objects'?

To look at this question, we first accept that distinct identifiable particles can exist. Imagine three of them. There are two possible realities. In the Machian view, the properties of the system are exhausted by the masses of the particles and their separations, but the separations are mutual properties. Apart from the masses, the particles have no attributes that are exclusively their own. They – in the form of a triangle – are a single thing. In the Newtonian view, the particles exist in absolute space and time. These external elements lend the particles attributes – position, momentum, angular momentum – denied in the Machian view. The particles become three things. Absolute space and time are an essential part of atomism.

The lent properties are the building blocks of both classical and quantum mechanics. Classically, each particle has a unique set of them, defining the state of each particle at any instant. This is the ideal to which realists like Einstein aspire. The lent properties also occur in quantum mechanics. They are generally not the state itself, but superpositions of them are. If a quantum system is considered in isolation from the instruments used to study it, its basic elements still derive from a Newtonian ontology. This is what misled EPR into thinking they could outwit Bohr. Einstein's defeat by Bohr is a clear hint that we shall only understand quantum mechanics when we comprehend Mach's 'overpowering unity of the All'.

BELL'S INEQUALITIES

Strong confirmation for quantum mechanics being holistic in a very deep sense was obtained in the 1960s, when John Bell, a British physicist from

Belfast, achieved a significant sharpening of the EPR paradox. The essence of the original paradox is the existence of correlations between pairs of quantities – pairs of positions or pairs of momenta – that are always verified if one correlation or the other is tested. By itself, some degree of correlation between the two particles is not mysterious. The EPR-type correlated states are generally created from known uncorrelated states of two particles that are then allowed to interact. Even in classical physics, interaction under such circumstances is bound to lead to correlations. Bell posed a sharper question than EPR: is the extent of the quantum correlations compatible with the idea that, before any measurement is made, the system being considered already possesses all the definite properties that could be established by all the measurements that, when performed separately, always lead to a definite result?

Bell's question perfectly reflects Einstein's 'robust realism' – that the two-particle system ought to consist of two separate entities that possess definite properties before any measurements are made. Assuming this, Bell proceeded to derive certain inequalities, justly famous, that impose upper limits on the degree of the correlations that such 'classical' entities could exhibit (tighter correlations would simply be a logical impossibility). He also showed that quantum mechanics can violate these inequalities: the quantum world can be more tightly correlated than any conceivable 'classical world'. Aspect's experiments specifically tested the Bell inequalities and triumphantly confirmed the quantum predictions. The only way in which the atomized world after which Einstein hankered can be saved is by a physical interaction that has so far completely escaped detection and is, moreover, propagated faster than light. Einstein could hardly have taken comfort from this straw. Far better, it seems to me, is to seek understanding of the Here in Mach's All. I shall give some indication of what I mean by this after we have considered the next topic.

THE MANY-WORLDS INTERPRETATION

In 1957, Hugh Everett, a student of John Wheeler at Princeton, proposed a novel interpretation of quantum mechanics. Its implications are startling, but for over a decade it attracted little interest until Bryce DeWitt drew wide attention to it, especially by his coinage *many worlds* to describe the main idea. Everett had used the sober title 'Relative state formulation of quantum mechanics'. One well-known physicist was prompted to call it the 'best-kept secret in physics'. So far as I know, Everett published no other scientific paper. He was already working for the Weapons Systems Evaluation Group at the Pentagon when his paper was published. He was apparently a chain smoker, and died in his early fifties.

Everett noted that in quantum mechanics 'there are two fundamentally different ways in which the state function can change': through continuous causal evolution and through the notorious collapse at a measurement. He aimed to eliminate this dichotomy, and show that the very phenomenon that collapse had been introduced to explain – our invariable observation of only one of many different possibilities that quantum mechanics seems to allow – is actually predicted by pure wave mechanics. Collapse is redundant.

The basis of Everett's interpretation is the endemic phenomenon of entanglement. By its very nature, entanglement can arise only in composite systems – those that consist of two or more parts. In fact, an essential element of the many-worlds interpretation as it is now almost universally understood is that the universe can and must be divided into at least two parts – an observing part and an observed part. However, Everett himself looked forward to the application of his ideas in the context of unified field theories, 'where there is no question of ever isolating observers and object systems. They are all represented in a *single* structure, the field.' That is ultimately the kind of situation that we must consider, but for the moment we shall look at the familiar form of the interpretation.

The simplest two-particle system can be used to explain a quantum measurement. The core idea is all that counts. One particle, called the *pointer*, is used to establish the location of the other particle, called the *object*. Figure 43 shows things with which we are already familiar. At an initial time t_{set-up}, the pointer (horizontal axis) and object (vertical axis) are not entangled. For any of the small range of possible pointer positions, the object has identical ranges of possible positions, shown schematically by points 1 to 6 on the left. Determination of the pointer position in this state would tell us nothing about the object. But the interactions of the particles are so arranged that by the later time $t_{measurement}$ the wave function, passing through the interaction region, has 'swung round' into the position shown on the right. Remembering how points in

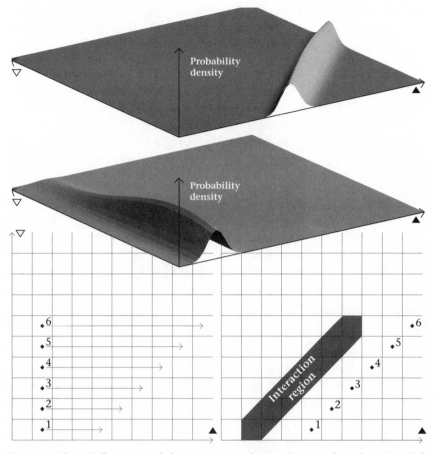

Figure 43 The initially unentangled state at $t_{\text{set-up}}$, shown schematically at the bottom left by the vertical column of positions 1, 2, 3, 4, 5, 6, and the first probability density above it, evolves into the entangled state at $t_{\text{measurement}}$, indicated on the right by the inclined numbers 1, 2, 3, 4, 5, 6 and the upper right probability density.

Q translate into positions in space, we see that the object still has its original range of positions 1 to 6, but that the new pointer positions are strongly correlated with them.

If we operate detectors at $t_{\text{measurement}}$ to find the pointer's position (by letting it strike an emulsion or even, in principle, by acting ourselves as observers and placing our eyes at the appropriate places), the standard quantum rules immediately tell us the object's position, for different pointer positions are now correlated with different object positions. Quantum measurement consists of two stages: the creation from an unentangled state of a strongly entangled state (creating the conditions of a so-called *good measurement*) followed by the exploitation of the correlations in that state (using the determination of the pointer position to deduce – measure – the object's position). The existence of such correlations has now been wonderfully well confirmed by experiments. If detectors have been used to find the pointer and are then used to locate the object directly, the quantum correlations predicted are invariably confirmed. Measurement theory is really verification of the correlations associated with entanglement. Personally, I think that the term 'measurement' has generated misunderstanding, and that it would be better simply to speak of verification of correlations.

The *measurement problem* of quantum mechanics is this: how does the entangled state of many possibilities collapse down to just one, and when does it happen? Is it when the pointer strikes the emulsion, or when the human observer sees a mark on the emulsion? I won't go into all the complications, which depend on how much of the world we wish to describe quantum mechanically. It leads to a vicious infinite regress. You can go on asking quantum mechanics again and again to say when collapse occurs, but it never gives an answer. The different possibilities already represented at $t_{\text{set-up}}$ in the different positions of the object system can never be eradicated, and simply 'infect' the rest of the world – first the pointer, then the emulsion, then the retina of the experimentalist's eye, finally his or her conscious state. All that the Copenhagen interpretation can say is that collapse occurs at the latest in the perceptions of the experimentalist. When it happens no one can say – it can only be said that if collapse does not happen we cannot explain the observed phenomena.

But must it happen? Everett came up with a simple – with hindsight obvious – alternative. Collapse does not happen at all: the multiple possibilities represented in the entangled state continue to coexist. In each possibility the observer, in different incarnations, sees something different, but what is seen is definite in each case. Each incarnation of the observer sees one of the possible outcomes that the Copenhagen interpretation assumes is created by collapse. The implications of this are startling. A single atomic particle – the object particle in Figure 43 – can, by becoming entangled with first the pointer and then the emulsion, and finally the conscious observer, split that observer (indeed the universe) into many different incarnations. In his paper in 1970 that at last brought Everett's idea to wide notice, Bryce DeWitt wrote:

I still recall vividly the shock I experienced on first encountering this multiworld concept. The idea of 10^{100+} slightly imperfect copies of oneself all constantly splitting into further copies, which ultimately become unrecognizable, is not easy to reconcile with common sense. Here is schizophrenia with a vengeance.

Everett's proposal raises two questions. If many worlds do exist, why do we see only one and not all? Why do we not feel the world splitting? Everett answered both by an important property of quantum mechanics called *linearity*, or the *superposition principle*. It means that two processes can take place simultaneously without affecting each other. Consider, for example, Young's explanation of interference between two wave sources. Each source, when active alone, gives rise to a certain wave pattern. If both sources are active, the processes they generate could disturb each other drastically. But this does not happen. The wave pattern when both sources are active is found simply by adding the two wave patterns together. The total effect is very different from either of the individual processes, but in a real sense each continues unaffected by the presence of the other. This is by no means always the case; in so-called non-linear wave processes, the wave pattern from two or more sources cannot be found by simple addition of the patterns from the separate sources acting alone. However, quantum mechanics is linear, so the much simpler situation occurs.

As a result, quantum processes can be regarded as being made up of many individual subprocesses taking place independently of one another. In Figure 43 the total process of the wave function 'swinging round' and becoming entangled is represented symbolically by the arrows as six individual subprocesses (or *branches*, to use Everett's terminology). In all of them, the pointer starts in the same position but ends in a different position. Everett makes the key assumption that conscious awareness is always associated with the branches, not the process as a whole. Each subprocess is, so to speak, aware only of itself. There is a beautiful logic to this, since each subprocess is fully described by the quantum laws. There is nothing within the branch as such to indicate that it alone does not constitute the entire history of the universe. It carries on in blithe ignorance of the other branches, which are 'parallel worlds' of which it sees nothing. The branches can nevertheless be very complicated. An impressive part of Everett's paper demonstrates how an observer (modelled by an inanimate computer) within one such branch could well have the experience of being all alone in such a multiworld, doing quantum experiments and finding that the quantum statistical predictions are verified.

Any scientific theory must establish a postulate of *psychophysical parallelism*: it is necessary to say what elements of the physical theory correspond to actual conscious experience. Given our current meagre understanding of consciousness, we have considerable freedom in the choice we make. Everett exploited the linearity of quantum mechanics to make his particular choice. It leads, however, to what is now widely seen as a serious technical problem.

Right at the start, Everett stated that 'The wave function is taken as the basic physical entity with no a priori interpretation.' He aimed to show that the interpretation of the theory emerges from 'an investigation of the logical structure of the theory'. This aim, coupled with his insistence that the wave function is the only thing that exists, creates the difficulty, since the logical structure of the theory is generally reckoned to be represented by Dirac's transformation theory. According to it, any quantum state can indeed be regarded as made up of other states – branches in an Everett-type 'many-worlds' picture. The difficulty is that this representation is not unique. There are many different ways in which one and the same state, formed from the same two 'observer' and 'object' systems, can be represented as being made up of other states. We can, for example, use position states, but we can equally well use momentum states.

The fact is that quantum mechanics is doubly indefinite. First, if states of a definite kind are chosen, any state of a composite system is a unique sum of states of its subsystems. For position states, this is shown in Figures 40 to 43. The probability distribution is spread out over a huge range of possibilities in which one particle has one definite position and the other particle has another definite position. Positions are always paired together in this way. Everett resolved the apparent conflict between our experience of a unique world and this multiplicity of possibilities by associating a separate and autonomous experience with each. However, he did not address the second indefiniteness: the states shown as positions in Figures 40 to 43 could equally well be represented by, for example, momentum states. Then pairs of momentum states result. Depending on the representation, different sets of parallel worlds are obtained: 'position histories' in the one case, 'momentum histories' in the other. One quantum evolution yields not only many histories but also many families of different kinds of history.

It was surprisingly long before this difficulty was clearly recognized as the *preferred-basis problem*: a definite kind of history will be obtained only if there exists some distinguished, or *preferred*, choice of the basis, by which is meant the kind of states used in the representation. The preferred basis problem is the EPR paradox in a different guise. Everett may have instinctively assumed that the position basis is somehow naturally singled out, but there is little evidence in his paper to confirm this.

The first question that must be addressed is surely this: what is real? Everett took the wave function to be the only physical entity. The price for this wave-function monism is the preferred-basis problem. Because the wave functions of composite systems can be represented in so many ways, the application of Everett's ideas to different kinds of representation suggests that one and the same wave function contains not only many histories, but also many different kinds of history. It leads to a 'many-many-worlds' interpretation. Some accept this, but I feel there is a more attractive alternative.

A DUALISTIC PICTURE

The purists among the quantum 'founding fathers', above all Dirac and Heisenberg, saw a close parallel between the representation of one and the same quantum state in many ways and the possibility of putting many different coordinate systems on one and the same space-time. In relativity, this corresponds to splitting space-time into space and time in different ways. After Einstein's great triumph, no physicist would dream of saying that this could be done in one way only. Similarly, Dirac and Heisenberg argued, there is nothing in quantum theory to suggest that there is a preferred way to represent quantum states. However, the parallel may not be accurate.

First, in classical relativity, space-time represents all reality – the complete universe. In contrast, a quantum state by itself has no definite meaning until the strategic decision – say, to measure position or momentum – has been taken. The state acquires its full meaning only in conjunction with actual measuring apparatus outside the system. The system must interact with the apparatus to reveal its latent potentialities. At present, its interaction with an apparatus – essentially the rest of the universe – is not fully understood. The quantum state by itself is only part of the story. It may be premature to draw conclusions about the quantum universe from incomplete quantum descriptions of subsystems of it.

Second, quantum mechanics as presently formulated needs an external framework. Indeed, the most basic observables, those for position, momentum and angular momentum, all correspond to the 'lent' properties mentioned in the discussion of the EPR paradox. They could not exist without the framework of absolute space, and Mach's principle suggests strongly it is determined by the instantaneous configurations of the universe. Time, moreover, plays an essential role in quantum mechanics yet stands quite outside the description of the quantum state. But we saw in Chapter 6 that time is really just a shorthand for the position of everything in the universe, so the configurations of the universe can be expected to play an essential and direct role in a quantum description of the universe. I cannot see how we can hope to understand the external framework of current quantum theory unless we put them into the foundations of quantum cosmology. This is what leads me to the dualistic picture of Platonia, the collection of all possible configurations of the universe, and the completely different wave function, conceived of as 'mist' over Platonia. In the language of Everett's theory, this introduces a preferred basis. In answer to the question 'what is real?', I answer 'configurations'. My book is the attempt to show that they explain both time and the quantum – as different sides of the same coin.

CHAPTER 15

The Rules of Creation

THE END OF CHANGE

In this chapter I am going to go into a little detail about how wave mechanics works. This means looking at two equations Schrödinger discovered in 1926 which, Dirac remarked, explained all of chemistry and most of physics. You will need to absorb enough to understand the bearing of the first part of the book on the structure of quantum cosmology. That is the goal; I hope you will find it is worth the effort. I believe it will show us how creation works. No theory can ever explain why anything is – that is the supreme mystery. But theory may be able to tell us why one thing rather than another is created and experienced. What is more, I believe that in every instant we experience creation directly. Creation did not happen in a Big Bang. Creation is here and now, and we can understand the rules that govern it. Schrödinger thought he had found the secret of the quantum prescriptions. Properly understood, what he found were the rules of creation.

Let us get down to business. We shall be considering how the wave function ψ changes. In quantum mechanics, this is all that does change. Forget any idea about the particles themselves moving. The space Q of possible configurations, or structures, is given once and for all: it is a timeless configuration space. The instantaneous position of the system is one point of its Q. Evolution in classical Newtonian mechanics is like a bright spot moving, as time passes, over the landscape of Q. I have argued that this is the wrong way to think about time. There is neither a passing time nor a moving spot, just a timeless path through the landscape, the track taken by the moving spot in the fiction in which there is time.

In quantum mechanics with time, which we are considering now, there is no track at all. Instead, Q is covered by the mists I have been using to illustrate the notion of wave functions and the probabilities associated with them. The red and green mists evolve in a tightly interlocked fashion, while the blue mist, calculated from the other two, describes the change of the probability. All that happens as time passes is that the patterns of mist change. The mists come and go, changing constantly over a landscape that itself never changes.

One of the equations that Schrödinger found governs this process. If ψ is known everywhere in Q at a certain time, you know what ψ will be slightly later. From this new value, you can go on another small step in time, and another, and so on arbitrarily far into the future. The role played here by the red and green mists, the two primary components of ψ, is quite interesting: the way the red mist varies in space determines the rate of change of the green mist in time, and vice versa. The two components play a kind of tennis. This equation is sometimes called the *time-dependent* Schrödinger equation because time features in it. This is not in fact the first equation that Schrödinger discovered.

The first one he found is now usually called the *stationary* or *time-independent* Schrödinger equation. This determines what happens in certain special cases in which the two components of ψ, the red and green 'mists', oscillate regularly, the increase of one matching the decrease of the other. This has the consequence, as we have already seen for a momentum eigenstate, that the blue mist (the probability density) has a frozen value – it is independent of time (though its value generally changes over Q). Such a state is called a *stationary state*. This explains the name given to the second equation – its solutions are stationary states. The standard view is that the time-dependent equation is the fundamental equation of quantum mechanics; the stationary equation is seen as a special case derived from it. This corresponds to an overall scheme in which some state of ψ is created at some time and then evolves until a measurement is made.

There are intriguing hints that in the quantum mechanics of the universe the roles of these two equations are reversed. The stationary equation (or something like it) may be the fundamental equation, from which the time-dependent equation is derived only as an approximation. We think it is fundamental because we have been fooled by circumstances that make it valid for the description of the phenomena we find around us. However, these phenomena deceive us greatly when it comes to the overall story of the universe. In particular, they lead us to believe time exists when it does not.

That this is likely to be so follows from an important property of the two Schrödinger equations. For any quantum system, we can use the time-independent equation to find all the stationary states it can have. Each of these states corresponds to a definite energy, and in each of them the red and green mists oscillate with the same fixed frequency while the blue mist remains constant. These solutions are also solutions of the time-dependent equation, though they are special, being stationary. I have mentioned linearity in quantum mechanics. Here, linearity means that two or more solutions of the time-dependent Schrödinger equation can be simply added together to give another solution. If the special stationary solutions are added, something significant results. In each solution, considered separately, the red and green mists oscillate at a fixed frequency while the blue mist remains constant. However, when we add two such solutions with different frequencies, they interfere: the added intensities of the red and green mists no longer oscillate regularly. More significantly, the blue mist varies in time.

Now this is very characteristic – indeed, it is the essence of quantum evolution. All solutions of the time-dependent equation can be found by adding stationary solutions with different frequencies. Each stationary solution on its own has regular oscillations of its red and green mists, but a constant – in fact static – distribution of its blue mist. But as soon as stationary states with different energies, and hence frequencies, are added together, irregular oscillations commence – in particular in the blue mist, the touchstone of true change. All true change in quantum mechanics comes from interference between stationary states with different energies. *In a system described by a stationary state, no change takes place.*

The italics are called for. We have reached the critical point. The suggestion is that the universe as a whole is described by a single, stationary, indeed static state. Why should this – with its implication that nothing happens – be so? This is where we start to make contact with the earlier part of the book. Time and change come to an end when Machian classical dynamics meets quantum mechanics. We have seen that a Machian universe should have only one value of the energy: zero. We also know (Box 2) that a quantum theory can be obtained by quantizing a corresponding classical theory. In fact, it is easy to show that whereas quantizing Newtonian dynamics, with its external framework of space and time, leads to the time-dependent Schrödinger equation, quantizing the simple Machian model considered in Chapter 7 leads to a quantum theory in which the basic equation is not the time-dependent but the stationary Schrödinger equation.

If the Machian approach to classical dynamics is correct, quantum cosmology will have no dynamics. It will be timeless. It must also be frameless.

CREATION AND THE SCHRÖDINGER EQUATION

Before I can explain how this can be achieved, I must tell you what the Schrödinger equation is like and what it can do. I believe it is even more remarkable than physicists realize. This is where – if I am right – we are getting near the secret of creation.

When Schrödinger created wave mechanics, Bohr's was the only existing model of the atom. It suggested that atoms could exist in stationary states, each with a fixed energy, photons being emitted when the atom jumped between them. Schrödinger's great aim was to explain how the stationary states arise and the jumps occur. De Broglie's proposal suggested strongly that a stationary state should be described by a wave function that oscillated rapidly in time with fixed frequency, though its amplitude might vary in space. As a first step Schrödinger therefore looked for an equation for the variation in space.

It is ironic that only later did he find the time-dependent equation from which, strictly speaking, he should have derived this equation. However, he had luck and was guided by good intuition. Although it is easy for mathematicians, I shall not go into the details of how Schrödinger found his equations or how to get from one to the other. Box 13 gives the minimum about the stationary equation needed to understand the thrust of the story.

BOX 13   How Creation Works

You can think of the Schrödinger wave function in a stationary state as follows. At each point of the configuration space Q, imagine a child swinging a ball in a vertical circle on a string of length ϕ, which remains constant. As the ball whirls, its height above or below the centre of the circle changes continuously. The height is an image of the red mist, which is sometimes positive (above the centre), sometimes negative (below it). The distance sideways – to the right (positive) or the left (negative) – is an image of the green mist. The square of ϕ is the image of the constant intensity of the blue mist. A stationary state is like having children swinging such balls at the same rate everywhere in Q, all perfectly in phase – they all reach the top of the circle together. The only thing not perfectly uniform is the string length, ϕ, which can change from point to point in Q. In a momentum eigenstate, ϕ is the same everywhere. It is a very special state, but in a more general stationary state ϕ does vary over Q. The stationary Schrödinger equation governs its variation.

It does this by imposing a condition at each point of Q. The sum of two numbers, calculated in definite ways, must equal a third. The first number is the most interesting but the most difficult to find. Take a quantum system of three bodies. Its configuration space Q has nine dimensions. Each point in Q corresponds to a position of the three bodies in absolute space. Imagine holding two bodies fixed, and moving the third along a line in absolute space. This will move you along a line in Q. Suppose that along it you plot ϕ, the string length, as a curve above the line. At each point, this curve will have a certain curvature. At some places it will curve strongly, towards or away from the line, at others weakly. In the calculus, the curvature is the second derivative.

At each point of Q there are nine such curvatures because Q has nine dimensions, one for each of the three directions in absolute space in which each particle can move. The first number in the Schrödinger condition is the sum of these nine curvatures after each has been multiplied by the mass of the particle for which it has been calculated. I shall call this the *curvature number*.

The second number is much easier to find. Recall that any configuration of bodies has an associated potential energy. The configuration (and the nature of the bodies, their masses, etc.) determines it uniquely. For gravity, this was explained in Figure 17. The second number, which I shall call the *potential number*, is found simply by multiplying the potential by ϕ.

The third number is also easy to find. If ω is the frequency of the state (the number of 'rotations of the balls' in a second), then, by the quantum rules, the energy of the state is $E = h\omega$, where h is Planck's constant. This is the relationship Einstein found between the energy and frequency of a photon. The third number, which I shall call the *energy number*, is then found by multiplying the energy E by ϕ.

The condition imposed by the stationary Schrödinger equation is then

Curvature number + Potential number = Energy number

(Planck's constant also occurs in the first number, to ensure that all three numbers have the same physical nature.)

However, finding this condition, which must hold everywhere in Q, was only half the story. Schrödinger thought that an atom in a stationary state was like a violin string vibrating in resonance. Because its two ends are fixed, the amplitude at the ends is zero. He therefore imposed on φ not only the above condition, but also the condition that it should tend to zero at large distances. It was this requirement that enabled him to make the huge discovery that convinced him – and very soon everyone else – that he had found the secret of Bohr's quantum prescriptions.

This hinges on an extremely interesting property of the stationary Schrödinger equation. As yet E is a fixed but unknown number. It may be smaller or greater than the potential V, which varies over Q. The interesting thing is that the above condition forces ϕ to do very different things depending on the value of $E - V$. Where it is greater than zero, ϕ oscillates. As Schrödinger said rather quaintly, 'it does not get out of control'. However, where $E - V$ is less than zero, the condition forces an entirely different behaviour on ϕ. It must either tend rapidly to zero or else grow rapidly – exponentially in fact – to infinity. The latter would be a disaster. Schrödinger therefore commented that things become tricky and must be handled delicately. Indeed, he showed that it is only in exceptional cases, for special values of E, that ϕ does not 'explode' but instead subsides to zero at infinity. These are the cases he was looking for. Well-behaved solutions exist for only certain values of E, which are discrete (separated from each other) if E is less than zero.

The well-behaved solutions are called *eigenfunctions*, and the corresponding values of E are called (energy) *eigenvalues*. It is a fundamental property of quantum mechanics that any system always has at least one eigenfunction. The eigenfunction of any system that has the lowest value of its energy eigenvalue (there is often only one such eigenfunction) is called the *ground state*. In general, there are also eigenfunctions with higher energies, called *excited states*. Finally, if E is large enough for $E - V$ to be positive everywhere, the eigenfunctions oscillate everywhere, though more rapidly where the potential is lowest. The negative eigenvalues E form the *discrete spectrum*, and the corresponding states are called *bound states* because for them ϕ has an appreciable value only over a finite region. The remaining states, with E greater than zero, are called *unbound states*, and their energy eigenvalues form the *continuum spectrum*.

Schrödinger won the 1933 Nobel Prize for Physics mainly for his wave-mechanical calculation for the hydrogen atom. He found that the energy eigenvalues of its stationary states are precisely the energies of the allowed states in Bohr's model. This was a huge advance, since Schrödinger's formalism had an inner unity and consistency to it completely lacking in the older model. Brilliant successes of the new wave mechanics, many achieved by Schrödinger himself, soon came flooding in, leaving no doubt about the great fruitfulness of the new scheme.

In Chapter 14 I described how molecules appear in the Schrödinger picture: as immense collections of all the configurations they could conceivably have, with the blue mist of the quantum probability strongly concentrated on the most probable configurations. These most probable configurations, generally clustered around a single point in Q, are the ones represented by the ball-and-strut models. I can now begin to make good my claim that Schrödinger found the laws of creation. His stationary equation determines the structures – indeed, creates the structures – of all these amazing atoms and molecules that constitute so much of the matter in the universe, our own bodies included. The equation does it by determining which structures are probable. But I mean creation not only in this sense of the structure of atoms and molecules, but in an even deeper one. The full explanation is still to come, but we are getting closer to our quarry.

QUANTUM MECHANICS HOVERING IN NOTHING

We must now see if we can dispense not only with time but also with absolute space in quantum mechanics. In a timeless system the energy E is zero, and the condition in Box 13 says simply that at every point of Q the sum of the curvature number and the potential number is zero. The potential number is already in the form we need. For any possible relative configuration, the potential has a unique value: it depends on nothing else. To find the potential number, we simply calculate the potential V for each configuration and then multiply by ϕ, getting $V\phi$. This part of the calculations is pleasingly self-contained because V depends only on the relative configuration. Each structure has its own potential irrespective of how we imagine the structure to be embedded in space.

However, a lack of 'self-containment' shows up in the curvature number. To find it, we must know how ϕ varies from position to position in the configuration space Q. This is not a self-contained process in Schrödinger's equation because the points of his Q are defined by the particles' positions in absolute space, which is used crucially in Q, making it *hybrid*. The all-important curvatures of ϕ are ultimately determined by position differences in absolute space. As a result, in standard quantum mechanics the orientations are in general entangled with the relative data that specify the particle separations. Now, besides positions, momenta and energy there is another very important quantity in quantum mechanics – angular momentum, which, being an action, always has discrete eigenvalues. It owes its existence in quantum mechanics to absolute space. We have not yet escaped from Newton's framework.

We are now coming to another critical point. We have seen that in classical physics the action is a kind of 'distance' between two configurations that are nearly but not exactly the same. Absolute space is an auxiliary device that makes it possible to define such 'distances'. This is why angular momentum exists in classical and quantum physics. However, in Chapter 7 we found an alternative definition of 'distances' that works in the purely relative configuration space – in Platonia – and owes nothing to absolute space. They are defined by the best-matching procedure, which uses relative configurations and nothing else. In classical physics, this makes it possible to create a purely relative and hence self-contained dynamics. We also found that a sophisticated form of best matching lies at the heart of general relativity. Best matching would appear to be a basic rule of the world.

It is therefore very tempting to see whether it can be applied in quantum mechanics. What we would like to do is establish rules for operating on wave functions defined solely on the relative configuration space. For example, for three bodies we would want to eliminate the six dimensions associated with their position and orientation in absolute space, and work just with the sides of the triangle. We shall then have a wave function defined on a three-dimensional Platonia. For that, we shall want to calculate a curvature number and a potential number. The latter will present no difficulty, since it will be the same as in ordinary quantum mechanics. The difficulty is in the curvature number. What, after all, is curvature? For any given curve, it is the rate at which its slope changes. But the key thing about a rate of change is that it is with respect to *something*. That something is all-important. It is a kind of 'distance'. The ordinary quantum-mechanical 'distance' is simply distance in absolute space (times the mass of the particle considered). To eliminate absolute space in classical physics, we replaced it by the Machian best-matching distance. There is no reason why we should not do the same in quantum physics.

This is where the unfolding of quantum mechanics on configuration space is so important. To retain that essential property of it – the huge step that Schrödinger took – we must pass from his hybrid Q to Platonia. If we are to succeed in formulating quantum mechanics in the new arena, there must be 'distances' in it. But that is precisely what the best-matching idea was developed to provide. Exactly the same 'distances' needed to realize Mach's principle in classical physics can be used in a version of wave mechanics for a universe without absolute space. All we have to do is measure curvatures with respect to the Machian distances created on Platonia by best matching. We then add curvatures measured in as many mutually perpendicular directions as there are dimensions in that time-less arena, and set the sum equal to minus the potential number.

In fact, it is quite easy to see that the wave functions that satisfy the Schrödinger conditions in this Machian case are precisely the eigenfunctions of ordinary quantum mechanics for which the angular-momentum eigenvalues are zero. This exactly matches our result in classical mechanics – that the best-matching condition leads to solutions identical to the Newtonian solutions with angular momentum zero. We have already seen why they must be static solutions.

The picture that emerges is very simple. The quantum counterpart of Machian classical dynamics is a static wave function Ψ on Platonia. The rules that govern its variation from point to point in Platonia involve only the potential and the best-matching 'distance'. Both are 'topographic features' of the timeless arena. Surveyors sent to map it would find them. They would see that the mists of Platonia respect its topography. It determines where the mist collects.

CHAPTER 16

'That Damned Equation'

HISTORY AND QUANTUM COSMOLOGY

The year 1980 was another turning point in my life. It was when Bruno Bertotti and I thought we might have found a new theory of gravitation, only to learn that the two ideas on which we had based it were already an integral part of Einstein's theory. Karel Kuchař's intervention rounded off our work but also brought it to an end. It was something of an anticlimax. Bruno became increasingly involved in experiments using spacecraft, aimed at detecting the gravitational waves predicted by Einstein's theory. For a year or two I actually stopped doing physics and became politically active in the newly founded Social Democratic Party (the SDP). However, the old interests soon revived. Margaret Thatcher's decisive general election victory in 1983 hastened the process.

Two things occupied me through the 1980s. First, I wrote the book from which I quoted the comments about Kepler. It had always been my ambition to write about absolute and relative motion, and in 1984 I signed a contract with Cambridge University Press for a book of four hundred pages covering the period from Newton to Einstein and including an account of my work with Bruno. When I embarked upon it, it occurred to me that I ought to find out why Newton had said what he had. What had given him the idea of absolute space? Might it not be an idea to look at what Galileo had said? I made a wonderful mistake by asking those questions. Before I knew what was happening, my research into Galileo dragged me ever further into past history, through the Copernican revolution to the work of Ptolemy and all the way back to the pre-Socratic philosophers. By reading the actual works of scientists such as Ptolemy, Kepler and Galileo, I found that the early history of mechanics and astronomy was far more interesting than any account of it I could find by the professional historians of science. They had missed all sorts of fascinating things, and their histories were quite inadequate. Inspired by Kepler's comment that the ways by which men discover things in the heavens are almost as interesting as the things themselves, I started to write about all the early work. I spent from 1985 to 1988 writing a completely unplanned book: *The Discovery of Dynamics*. My sympathetic and understanding editor at Cambridge, Simon Capelin, agreed to publish it as the first of a two-volume work. The second volume was to be the book originally proposed and should have been completed a year or two later. However, that got badly delayed by a parallel development that turned my interest to physics that does not yet exist at the same time as I was working backward to the early history.

As I mentioned earlier, Bruno and I had been completely concerned with classical physics. We had wanted to show that Mach had been right and that his ideas could lead to new classical physics; we had given not a moment's thought to any quantum implications they might have. Quantum cosmology was a world beyond our ken. It is strange what sparks a desire to work on something. My lack of interest in quantum gravity was particularly odd, since it was the early work done in that field which, through the remark by Dirac, quoted in the Preface, had set me on my long trek. It was the same work that had led to the work of Baierlein, Sharp and Wheeler that Bruno and I had come to see as the implementation of Mach's ideas within general relativity. Not even working with Karel Kuchař, one of the world's leading experts in quantum gravity, provided the stimulus I needed. Perhaps it all seemed too daunting. I needed the example and encouragement that came from a new friend, Lee Smolin.

I first met Lee a few weeks before I travelled to Salt Lake City in the autumn of 1980. It was quite a dramatic time for me since I had just narrowly escaped death through an insidious appendix that had burst without giving me any pain. My only symptoms were tiredness, slight sickness and the merest hint of stomach pain. Luckily my vigilant doctor sent me to hospital as a precaution. An X-ray proved difficult to interpret, and after quite lengthy deliberation the doctors decided to open me up. They found that any further delay could have been fatal. Seeing my state, the surgeon apparently commented that 'this must be a very brave man', believing I must have been in agony. In fact, I had been cheerfully reading *The Times* without any discomfort only half an hour before the operation. The day after I came back from hospital still convalescing, two American physicists visiting Oxford phoned to say that they had heard from Roger Penrose about my interest in Mach's principle. Could they come and see me? They came the next day, and I greeted them in my dressing gown.

One was Lee, then a young postdoc. The meeting changed both of our lives significantly. He proved very receptive to the ideas of Leibniz and Mach to which I introduced him, while he encouraged me to see what application they might have to the problem to which he had decided to devote himself – quantum gravity. We met several times in the next few years, and collaborated on an attempt to formulate Leibniz's philosophical system, his 'monadology', in mathematical form. I think we made some real progress. Lee has written about his view of things in his *The Life of the Cosmos*. Certain aspects of our work together were decisive in my own elaboration of the notion of time capsules and my conviction that the ultimate and only truly real things are the instants of time. As far as I am aware, Leibnizian ideas offer the only genuine alternative to Cartesian–Newtonian materialism which is capable of expression in mathematical form. What especially attracts me to them is the importance, indeed primary status, given to structure and distinguishing attributes, and the insistence that the world does not consist of infinitely many essentially identical things – atoms moving in space – but is in reality a collection of infinitely many things, each constructed according to a common principle yet all different from one another. Space and time emerge from the way in which these ultimate entities mirror each other. I feel sure that this idea has the potential to turn physics inside out – to make the interestingly structured appear probable rather than improbable. Before he became a poet, T. S. Eliot studied philosophy. He remarked, 'In Leibniz there are possibilities.'

In 1988, when I had finished my book on the discovery of dynamics, I spent three weeks with Lee at Yale, and began to think seriously how one might make sense of the embryonic form of quantum gravity that had been developed from about the time of Einstein's death in 1955, leading to the publication of the Wheeler–DeWitt equation in 1967. During the next four years, Lee and I had many discussions. Although we eventually followed different paths – Lee is reluctant to give up time as a primary element in physics – the ideas I want to describe in the final part of the book crystallized during those discussions. For me, their attraction stems from the inherent plausibility of Platonia as the arena of the universe and the implication of Schrödinger's breathtaking step into a rather similar configuration space. As I see it now, the issue is simple.

A SIMPLE-MINDED APPROACH

You can play different games in one and the same arena. You can also adjust the rules of a game as played in one arena so that it can be played in a different arena. Both general relativity and quantum mechanics are complex and highly developed theories. In the forms in which they were originally put forward, they seem to be incompatible. What I found to my surprise was that it does seem to be possible to marry the two in Platonia. The structures of both theories, stripped of their inessentials, mesh. What if Schrödinger, immediately after he had created wave mechanics, had returned to his Machian paper of only a year earlier and asked himself how Machian wave mechanics should be formulated? His Machian paper implicitly required Platonia to be the arena of the universe, while any wave mechanics simply had to be formulated on a configuration space. Such is Platonia, though it is not quite the hybrid Newtonian Q he had used. But the structure of Machian wave mechanics would surely have been immediately obvious to him, especially if he had taken to heart Mach's comments on time. As a summary of the previous chapter, here are the steps to Machian wave mechanics in their inevitable simplicity.

For a system of N particles, the Schrödinger wave function in the Newtonian case will in general change if the relative configuration is changed, if the position of its centre of mass is changed, if its orientation is changed, and if the time is changed. Mathematicians call these things the *arguments* of the wave function. They constitute its arena. To see what really counts, we can write the wave function in the symbolic way that mathematicians do:

$$\psi \text{ (relative configuration, centre of mass, orientation, time)} \qquad (1)$$

But if the N particles are the complete universe, there cannot be any variation with change of centre of mass, orientation or time for the simple reason that these things do not exist. The Machian wave function of the universe has to be simply

$$\Psi \text{ (relative configuration)} \tag{2}$$

Note the grander Ψ. This is the *wave function of the universe*. It has found its home in Platonia.

I have met distinguished theoretical physicists who complain of having tried to understand canonical quantum gravity, the formalism through which the Wheeler–DeWitt equation was found, and have given up, daunted by the formalism and its seemingly arcane complexity. But, as far as I can see, the most important part boils down simply to the passage from the hybrid (1) to the holistic (2).

'THAT DAMNED EQUATION'

This is a bold claim, but the fact is that it still remains the most straight-forward way to understand the Wheeler–DeWitt equation. To conclude Part 4, I shall say something about this remarkable equation and the manner of its conception, which unlike the hapless Tristram Shandy's was inevitable, being rooted in the structure of general relativity. You may find this section a little difficult, which is why I have just given the simple argument by which I arrive at its conclusion. Just read over any parts you find tough.

That there was a deep problem of time in a quantum description of gravity became apparent at the end of the 1950s in the work of Dirac and Arnowitt, Deser and Misner (ADM) described in Chapter 11. The exis-tence of the problem was – and still is – mainly attributed to general covariance. The argument goes as follows. The coordinates laid down on space-time are arbitrary. Since the coordinates include one used to label space-time in the time direction and all coordinates can be changed at whim, there is clearly no distinguished *label of time*. This is what leads to the plethora of paths when a single space-time is represented as histories in Platonia. However, the real root of the problem lies in the deep structure of general relativity that we considered in the same chapter.

Indeed, as Dirac and ADM got to grips with the dynamics of general relativity, the problem began to take on a more concrete shape. The first fact to emerge clearly was the nature of the 'things that change'. This was very important, since it is the 'things that change' that must be quantized. They turned out to be 3-spaces – everything in the universe on one simultaneity hypersurface, including the geometrical relationships that hold within it. These are the analogue of particle positions in elementary quantum mechanics. As I have mentioned, Dirac was quite startled by this discovery – it clearly surprised him that dynamics should distinguish three-dimensional structures in a theory of four-dimensional space-time. I am surprised how few theoreticians have taken on board Dirac's comments. Many carry on talking about the quantization of *space-time* rather than *space* (and the things within it). It is as if Dirac and ADM had never done their work. Theoreticians are loath to dismantle the space-time concept that Minkowski introduced. I am not suggesting anything that he did is wrong, but it may be necessary to accommodate his insight to the quantum world in unexpected ways. One way or another, something drastic must be done.

As explained in Chapter 11, in general relativity four-dimensional space-time is constructed out of three-dimensional spaces. It turns out that their geometry – the way in which they are curved – is described by three numbers at each point of space. This fact of there being three numbers acquired a significance for quantum gravity a bit like the Trinity has for devout Christians. Intriguingly, the issue at stake is somewhat similar – is this trinity one and indivisible? Is one member of the trinity different in nature from the other two? The reason why the three numbers at each space point turned into such an issue is because it seems to be in conflict with a fact of quantum theory that I need to explain briefly.

I mentioned in Chapter 12 the 'zoo' of quantum particles, which are excitations of associated fields. The typical example is the photon – the particle conjectured by Einstein and associated with Maxwell's electro-magnetic field. An important property of particles is rest mass. Some have it, others do not. The massless particles must travel at the speed of light – as the massless photon does. In contrast, electrons have mass and can travel at any speeds less than the speed of light.

Now, massless particles are described by fewer variables (numbers) than you might suppose. Quantum mechanically, a photon with mass would be associated with vibrations, or oscillations, in three directions: along the direction of its motion (longitudinal vibrations) and along two mutually perpendicular directions at right angles to it (transverse vibrations). However, for the massless photon the longitudinal vibrations are 'frozen out' by the effects of relativity, and the only physical vibrations are the two transverse ones. These are called the two *true degrees of freedom*. They correspond to the two independent polarizations of light. This remark may make these rather abstract things a bit more real for the non-physicist. Humans cannot register the polarization of light, but bees can and use it for orientation.

There are many similarities between Maxwell's theory of the electromagnetic field and Einstein's theory of space-time. During the 1950s this led several people – the American physicist Richard Feynman was the most famous, and he was followed by Steven Weinberg (another Nobel Laureate and author of *The First Three Minutes*) – to conjecture that, just as the electromagnetic field has its massless photon, the gravitational field must have an analogous massless particle, the *graviton*. It was automatically assumed that the graviton – and with it the gravitational field – would also have just two true degrees of freedom.

From 1955 to about 1970, much work was done along these lines in studies of a space-time which is almost flat and therefore very like Minkowski space (I did my own Ph.D. in this field). In this case, the parallel between Einstein's gravitational field and Maxwell's electromagnetic field becomes very close, and a moderately successful theory (experimental verification is at present out of the question, gravity being so weak) was constructed for it. Within this theory it is certainly possible to talk about gravitons; like photons, they have only two degrees of freedom. However, Dirac and ADM had set their sights on a significantly more ambitious goal – a quantum theory of gravity valid in all cases. Here things did not match up. The expected two true degrees of freedom did not tally with the three found from the analysis of general relativity as a dynamical theory – as geometrodynamics.

Within the purely classical theory, the origin of the mismatch is clear: it is the criss-cross best-matching construction of space-time that I illustrated with the help of Tristan and Isolde. However, the discrepancy between the quantum expectations of well-behaved massless particles with two polarizations and the intricate interstreaming reality of relativity rapidly became the central dilemma of quantum gravity. Forty years on, it has still not yet been resolved to everyone's satisfaction – it is that intractable. This is perhaps not surprising, for the issue at stake is the fabric of the world. Does it exist in something like that great invisible framework that took possession of Newton's imagination, or is the world self-supporting? Do we swim in nothing? Nobody has yet been able to make quantum theory function without a framework. In fact, many people do not realize that the framework is a potential problem – Dirac's transformation theory is in truth the story of acrobatics in a framework, and for physicists nurtured on Dirac's *The Principles of Quantum Mechanics* the acrobatics is quantum theory. Acrobatics must be precise – if the trapeze is not where it should be, death can result. Such are the exigencies that led the early researchers to posit a graviton with just two true degrees of freedom.

An intriguing way to achieve this was suggested by Baierlein, Sharp and Wheeler's paper in 1962 and its enigmatic hint that 'time' was somehow carried within space. This was taken literally, especially since it seemed to solve another problem of quantum acrobatics. In real acrobatics, not only location but also timing is of the essence. Nobody knew how to do quantum mechanics without an independent time external to the quantum degrees of freedom. But such a time appeared to have gone missing in gravity. Instead of time and two true degrees of freedom, there appeared to be no time but three degrees of freedom; these, moreover, were suspect. The count was all too suggestive – and many people came to the same conclusion: there is a time, but it is hidden in the three degrees of freedom.

According to this insight, the basic framework of quantum mechanics could be preserved, but the time it so urgently needed would be taken from the 'world' to which it was to be applied. Putting it in very figurative terms, one-third of space would become time, while the remaining two-thirds would become two true quantum degrees of freedom. Because time was to be extracted from space, from within the very thing that changes, the time that was to be found was called *intrinsic time*. The notion of intrinsic time was – and is – a breathtaking idea. But there was a price to be paid, and there was also a closely related problem to be overcome: which third of space is to be time?

The problem was that no clear choice could be made. Any and all 3-spaces can appear in the relations that summarize so beautifully the true essence of general relativity. What is more, any choice would ultimately amount to the introduction of distinguished coordinates on space-time.

But this would run counter to the whole spirit of relativity theory, the essence of which was seen to be the complete equivalence of all coordinates. So if a choice were made, the price would be the loss of this equivalence. The price and the problem are one and the same. They presented the quantum theoreticians with a head-on collision between the basic principles of their two most fundamental theories – the need for a definite time in quantum mechanics and the denial of a definite time in general relativity. At an international meeting on quantum gravity held at Oxford in 1980, Karel Kuchař, concluding his review of the subject, stated that the problem of 'quantum geometrodynamics is not a technical one, but a conceptual one. *It consists in the diametrically opposite ways in which relativity and quantum mechanics view the concept of time'*. I have added the italics. I was there to hear the talk, and Kuchař's comment made a deep impression on me.

The search for the third of space that would become time has been like *The Hunting of the Snark*, Lewis Carroll's mythical beast that no one could find. Since the idea of intrinsic time was first clearly formulated about thirty-five years ago, the beast has not been found. Karel has done more than anyone else to try to track it down. If he cannot find it, I feel that comes quite close to a non-existence proof. My own belief is that the idea is based on an incorrect notion of time. It is a mythical beast invoked in vain to solve a titanic struggle. It does not surprise me that a special time has not been found lurking in the tapestry of space-time. All I see in that tapestry are change and differences – and the differences are measured democratically. The idea of a special intrinsic time to be extracted out of space, or out of any part of space-time or its contents, violates the democratic theory of emphemeris time that lies at the heart of general relativity.

If we look at the Newtonian parallel of the notion, it seems strange. In a world of three particles, it is like saying that one of the sides of the triangle they form is time while the other two are true degrees of freedom. Such an attempt to find time breaks up the unity of the universe. No astronomer observing a triple-star system would begin to think like that. The key property of astronomical ephemeris time is that all change contributes to the measure of duration. There has to be a different way to think about time.

I believe it was found, perhaps unintentionally, by Bryce DeWitt in 1967. John Wheeler had strongly urged him to find the fundamental equation of quantum gravity. It was Wheeler's high priority to find the Schrödinger equation of geometrodynamics. What the theory of intrinsic time should yield is a time-dependent Schrödinger equation that – in figurative language – evolves a wave function for 'two-thirds of space' with respect to a 'time' constituted by the remaining 'one-third of space'. Balking at the invidious task of selecting which third should be 'time', DeWitt fell back on a very general formalism developed fifteen years earlier by Dirac that made it possible to avoid having to make a choice.

Dirac's method makes it possible to treat all parts of space on an equal footing, and simply defers to later the problem of time. DeWitt used Dirac's method to write the fascinating equation that, as Kuchař noted, he himself calls 'that damned equation', John Wheeler usually calls the 'Einstein–Schrödinger equation' and everyone else calls the 'Wheeler–DeWitt equation'. But what is this equation, and what does it tell us about the nature of time?

The most direct and naive interpretation is that it is a stationary Schrödinger equation for one fixed value (zero) of the energy of the universe. This, if true, is remarkable, for the Wheeler–DeWitt equation must, by its nature, be the fundamental equation of the universe. I pointed out in the discussion of the structure of molecules that the 'ball-and-strut' models are only approximations to the quantum description, being merely the most probable configurations. The Wheeler–DeWitt equation is telling us, in its most direct interpretation, that the universe in its entirety is like some huge molecule in a stationary state and that the different possible configurations of this 'monster molecule' *are the instants of time*. Quantum cosmology becomes the ultimate extension of the theory of atomic structure, and simultaneously subsumes time.

We can go on to ask what this tells us about time. The implications are as profound as they can be. Time does not exist. There is just the furniture of the world that we call instants of time. Something as final as this should not be seen as unexpected. I see it as the only simple and plausible outcome of the epic struggle between the basic principles of quantum mechanics and general relativity. For the one – in its standard form at least – needs a definite time, but the other denies it. How can theories with such diametrically opposed claims coexist peacefully? They are like children squabbling over a toy called time. Isn't the most effective way to resolve such squabbles to remove the toy? We have already seen that there is a well-defined sense in which classical general relativity is timeless. That is, I believe, the deepest truth that can be read from its magical tapestry. The question then is whether we can understand quantum mechanics and the existence of history without time. That is what the rest of the book is about.

History in the Timeless Universe

If things simply are, how can history be? If quantum cosmology is merely a static mist that enshrouds eternal Platonia, whence the manifest appearance of motion and our conviction history is real? This is the great question. I am not going to give any summary of Part 5: read on, please. The kingfisher just about to set off in flight is a symbol for the task. Explaining how we see it in motion in a timeless world is no more of a problem than explaining why we are convinced that Henry VIII had six wives.

CHAPTER 17

The Philosophy of Timelessness

You should by now recognize the connection between the picture that emerges from the simplest interpretation of the Wheeler–DeWitt equation and the timeless world I sketched in Part 1. I outlined there, using the notion of the time capsule, how the seemingly dead and static Platonia might correspond to the vibrant living world we experience in every instant. In this final part of the book, I want to explain the arguments from physics that led me to the notion of time capsules, and also to show that the structure of quantum cosmology may well cause the wave function of the universe to 'seek out' time capsules. This is the story of how physics brings the Platonic forms to life. I start with some general comments.

I believe in a timeless universe for the childlike reason that time cannot be seen – the emperor has no clothes. I believe that the universe is static and is described by something like the Wheeler–DeWitt equation. I would like you to accept this as a working hypothesis, so we can see where it leads. As I said earlier, I believe that it leads to the rules of creation. Let me now explain why.

According to many accounts, in both mainstream science and religion, the universe either has existed for ever or was created in the distant past. Creation in a primordial fireball is now orthodox science – the Big Bang. But why is it supposed that the universe was created in the past rather than newly created in every instant that is experienced? No two instants are identical. The things we find in one are not exactly the same as the things we find in another. What, then, is the justification for saying that something was created in the past and that its existence has continued into the present?

The most obvious reason is the apparent persistence of objects and living beings. If pressed, though, we acknowledge that they never remain exactly the same. Even rocks weather slowly. However, enough properties remain unchanged for us to say that the same things do continue to exist. Indeed, human existence is inconceivable without a significant degree of stability in the world. No doubt the baby's recognition of the continually reappearing smiling face of its mother soon implants the notion of persistence. But if we want to think rationally and as philosophers about these matters, we ought to cultivate a degree of detachment. We must practise Cartesian doubt and, just once at least, question all our preconceptions.

I am not persuaded that the people who ought to be best at this – theoretical physicists – do achieve full freedom of thought. Many are passionately committed to an objectively existing external world. They hate anything that smacks of solipsism or creationism. This explains the controversies, virulent at times, about the reality of atoms that took place a century ago, and the equally impassioned debates today about the meaning of quantum mechanics (in many ways a continuation of the debate about atoms). For scientists committed to realism, atoms that remain the same in themselves and merely move in space and time are very welcome. Atoms, space and time are the things that either existed for ever or else came into being with the Big Bang.

However, the fields introduced by Faraday and Maxwell now provide the basis of quantum field theory, which is currently the deepest known form of quantum theory, and such fields are in perpetual flux. And within classical physics Einstein made space and time equally fluid and transient. Today there is only one scientific justification for saying that the universe was created in the past: the hypothesis of lawful dynamical evolution from some past, into the present, and on into an as yet unexperienced future. If an initial state uniquely determines a subsequent state of the same generic kind which differs only in detail, it is reasonable to speak of initial creation and subsequent evolution.

But this view must be challenged. It belongs to a mindset that holds the world either to be classical in its entirety, or to have quantum objects within the old classical framework of space and time. How slow we are to move out of old quarters! All the evidence indicates that anything dynamical must obey the rules of quantum mechanics even if it appears classical to our senses. But Einstein made space dynamical – that is the lesson of geometrodynamics taught us in detail by Dirac; by Arnowitt,

Deser and Misner (ADM); and by Baierlein, Sharpe and Wheeler (BSW). When space submits to the quantum, as it surely must, the last vestige of a created but persisting framework is lost. Moreover, the transition from the classical world we see to the quantum world that underlies it is fixed in its broad outlines. All we need do is put together the two things that go into quantization – a classical theory and the rules to quantize it – and see what comes out.

The central insight is this. A classical theory that treats time in a Machian manner can allow the universe only one value of its energy. But then its quantum theory is singular – it can only have one energy eigenvalue. Since quantum dynamics of necessity has more than one energy eigenvalue, quantum dynamics of the universe is impossible. There can only be quantum statics. It's as simple as that!

In Part 1 I mentioned the dichotomy in physics between laws and initial conditions. Most equations in physics do not by themselves give complete information, they only put limits on what is possible. To arrive at some definite prediction, further conditions are necessary. Neither Newton's nor Einstein's equations tell us why the universe has its present form. They have to be augmented by information about a past state. We could invoke a deity in the way Einstein was wont, who goes through two steps in creating the universe. First, laws are chosen, then an initial condition is added. Many people have wondered whether this is a permanent condition of physics.

The stationary Schrödinger equation is quite different in this respect. It obviously cannot have initial conditions, since it is a timeless equation. It does not require boundary conditions, either. Let me explain what this means. There are many equations in physics which describe how quantities vary in space without there being any change in time. Such equations can have many different solutions, and to find the one that is applicable in a specific case, mathematicians often stipulate the actual values the solution must have at the boundary of some region. This stipulation is what is called a *boundary condition*. Boundary conditions have the same kind of importance as initial conditions. However, as explained in Box 13, the stationary Schrödinger equation requires no such conditions. Instead, there is just a general condition on the way the wave function behaves. It must be continuous (not make any jumps), it must have only one value at each point and it must remain finite everywhere. As we saw, the condition of remaining finite – of not rushing off to infinity – is very powerful. It was what unlocked the quantum treasure chest. In fact, the first two conditions are also very powerful and lead to many important results. To distinguish these conditions from normal initial or boundary conditions, let me call them *conditions of being well behaved*. Mathematicians may regard this as somewhat artificial, since the condition of remaining finite does actually enforce a definite kind of behaviour at boundaries. It is therefore in some sense equivalent to a boundary condition. However, I prefer not to think of it in that way, since it is very general and can be formulated in a completely timeless fashion. It avoids all particular specification, which must always be arbitrary.

Now, my suggestion is this. There are no laws of nature, just one law of the universe. There is no dichotomy in it – there is no distinction between the law and supplementary initial or boundary conditions. Just one, all-embracing static equation. We can call it the *universal equation*. Its solutions (which may be one or many) must merely be well behaved, in the sense explained in the previous paragraph. It is an equation that creates structure as a first principle, just as the ordinary stationary Schrödinger equation creates atomic and molecular structure. This is because it attaches a ranking – a greater or lesser probability – to each conceivable static configuration of the universe.

I explained in connection with Figure 40 how the density of the blue mist can be used to create a collection of configurations in a bag, a heap even, from which the most probable atomic configurations can be drawn at random. Configurations – which are structures – are created as more or less definite potentialities to the extent that the stationary Schrödinger equation tells us to put more or less into the heap. Like the individual structures within it, the heap is static. It is carefully laid up in a Platonic palace, which, since probabilities play such a mysterious role in quantum mechanics, is a kind of 'antechamber of Being'.

Now I can start to make good my deeper claims about Schrödinger and creation. We have to forget all previous physics and approach things with an open mind. First, we look at what the Machian time-independent Schrödinger equation is and what it does. It is completely self-contained. For a system of three bodies it just works on triangles and masses, and nothing else. In a timeless fashion, it associates a probability with each triangle. This is tantamount to giving them a ranking. It is particularly suggestive that this ranking is determined by the triangles themselves – nothing else is involved. The probabilities for the triangles emerge from a comprehensive testing and comparison programme. The equation 'looks' at all possible wave functions that could exist on Platonia and throws out all those that do not 'resonate' properly. Those that are left have to be finely tuned, otherwise they will satisfy neither the equation nor the condition of being well behaved. And it is not just the wave function that resonates. We can say that the triangles that get the greatest probability are the ones that 'resonate best with their peers', since the triangles alone determine how the probability is distributed. This is what the rationality of best matching in classical dynamics translates into in quantum cosmology. There is a perfect, circle-closing, rational explanation for all the relative probabilities.

I do believe that what we have here are putative rules of creation, or perhaps we should say of being. Considered purely as an intellectual exercise, this quantum-mechanical determination of probabilities for relative configurations is no odder than the classical-dynamical determination of curves in configuration space. The aim of science is to find rational and economic explanations of observed phenomena, not to prejudge the issue. Each hypothetical scheme should be judged on its merits. There should be a clear statement of the phenomena that are to be explained, the conceptual entities that are to be employed and the mechanism that is to yield the explanation.

The first aim is to create a realist (non-solipsistic) cosmology in which there are sentient beings whose primary awareness is of structured instants of time as defined earlier in the book. These instants are like subjective snapshots, and may be called atoms of perceptual existence. Each snapshot holds together in an indissoluble unity everything that we would want to call the actual facts of which we are aware in an instant of time. These include not only the things we see, feel and hear, but also our awareness of them, our memories and our interpretation of everything. The fact that many different things are known at once is regarded (by me at least) as the most remarkable – and defining – property of instants of time. I do not believe that science (or religion) will ever explain why we experience instants, but perhaps it can explain the structure we find within them.

The scheme is realist because the structure of an external, objectively existing real thing is being proposed as the explanation of the structure experienced within a perceptual instant. What we experience in subjective instants reflects, through psychophysical parallelism, physical structure in external things: configurations of the universe. Their actual nature is a matter for ongoing research. The notion has been illustrated by configurations of mass points in Euclidean space, by island-type distributions of fields of Faraday–Maxwell type in Euclidean space, and by closed Riemannian 3-geometries (which may also have fields defined on them). It is at the last level that I believe satisfactory explanations can in principle be obtained for many of the known facts of physics and cosmology. However, some further development, very possibly associated with the notions of superstrings and supersymmetry, may well be needed to explain the actual cocktail of forces and particles that pervades the universe.

What is important about relative configurations is that they are intrinsically defined – they are self-contained things – and that the rule that defines one thing simultaneously defines many. Moreover, they can all be arranged systematically in a relative configuration space: Platonia, as I have called it.

Classical physics before general relativity 'explained' the world by assuming it to be a four-dimensional history of such relative configurations located in a rigid external framework of absolute space and time. Such a world is supposed to have evolved from certain initial conditions to the state we now observe by means of the laws of classical dynamics, in which the framework of space and time play a significant role. These laws provide all the explanation of which classical physics is in principle capable. In Part 2 I showed how the external framework can be dispensed with. It does not need to be invoked to formulate the laws of dynamics, nor even to visualize how things are located in space and time. Schrödinger's Kantian appeal to space and time as the ineluctable forms of thought was unnecessary. We can form a clear conception of structured things that stand alone. We have seen how this is also true of general relativity, in which space-time is 'constructed' by fitting together 3-spaces in a very refined and sophisticated way.

So, then, what does the Wheeler–DeWitt equation tell us can happen in a rational universe? The answer is ironic. Nothing! The quantum universe just is. It is static. What a denouement. This is a message that needs to be shouted from the rooftops. But how can this seemingly bleak message reverberate around a static universe? How can we bring dead leaves to life? The poet Shelley called on the wild west wind to carry his thoughts over the universe. What can play the role of the wind in static quantum Platonia?

CHAPTER 18

Static Dynamics and Time Capsules

DYNAMICS WITHOUT DYNAMICS

DeWitt already clearly saw the problem posed at the end of the last chapter – the crass contradiction between a static quantum universe and our direct experience of time and motion – and hinted at its solution in 1967. Quantum correlations must do the job. Somehow they must bring the world alive. I shall not go into the details of DeWitt's arguments, since he saw them only as a first step. However, the key idea of all that follows is contained in his paper. It is that the static probability density obtained by solving the stationary Schrödinger equation for one fixed energy can exhibit the correlations expected in a world that does evolve – classically or quantum mechanically – in time. We can have the appearance of dynamics without any actual dynamics.

It may surprise you, but it was about fifteen years before physicists, and then only a few, started to take this idea seriously. The truth is that most scientists tend to work on concrete problems within well-established programmes: few can afford the luxury of trying to create a new way of looking at the universe. A particular problem in everything to do with quantum gravity is that direct experimental testing is at present quite impossible because the scales at which observable effects are expected are so small.

Something like a regular research programme to recover the appearance of time from a timeless world probably began with an influential paper by Don Page (a frequent collaborator of Stephen Hawking) and William Wootters in 1983. This was followed by several papers that concentrated on an obvious problem. In ordinary laboratory physics, the fundamental equation used to describe quantum phenomena is the time-dependent Schrödinger equation. It undoubtedly holds to an extraordinarily good accuracy for all ordinary physics: we could not even begin to understand, for example, the radiation of atoms without this equation. But if the universe as a whole is described by a stationary Schrödinger equation and time does not exist at all, how does a Schrödinger equation with time arise? This question seems to have been first addressed by the Russians V. Lapchinskii and V. Rubakov, but a paper in 1985 by the American Tom Banks did more to catch the imagination of physicists. This was followed in 1986 by a paper treating the same problem by Stephen Hawking and his student Jonathan Halliwell. Further papers on the subject appeared in the following years. The whole associated research programme has become known as the *semiclassical approach*, for a reason I shall explain later. The basic idea is easy to grasp.

Imagine yourself on a wide sandy beach on which the receding tide has left a static pattern of waves. As you are a free agent, nothing can stop you from laying out a rectangular grid on the beach and calling the direction along one axis 'space' and that along the perpendicular axis 'time'. For each value of the 'time coordinate', you can examine the wave pattern along the one-dimensional line of 'space' at that 'time'. When you move to the neighbouring line on the beach corresponding to 'space' at a slightly later 'time', you will find that the wave pattern has changed. Simply by laying out your grid and calling one direction 'space' and the other 'time', you have transformed – in your mind's eye – a two-dimensional static picture into wave dynamics in one dimension. This can be done with wave patterns in spaces of any dimension N. One direction can always be called 'time', and this automatically creates 'evolution' in the remaining $N - 1$ dimensions.

Of course, if the original wave pattern is 'choppy' and has not been created by some rule, the choice of the 'time' direction will be arbitrary. Any choice will create the impression of evolution in the remaining $N - 1$ dimensions, but it will not obey any definite and simple law. In the semiclassical approach, there are two decisive differences from the arbitrary situation. First, the static wave pattern is the solution of a definite equation. Second, it is a somewhat special solution – called a *semiclassical solution* – in that it exhibits a more or less regular wave pattern. This assumption will be considered later. However, if the wave pattern satisfies the assumption, it automatically selects a direction that it is natural to call *time*. With respect to this direction, a genuine appearance of dynamics arises in a static situation (Box 14). The result is this. Two static wave patterns (in a space of arbitrarily many dimensions) can, under the appropriate conditions, be interpreted as an evolution in time of the kind expected in accordance with the time-dependent Schrödinger equation. The appearance of time and evolution can arise from timelessness.

BOX 14   The Semiclassical Approach

This box provides some necessary details about the semiclassical approach. It is important here that the quantum wave function is not one wave pattern but two (the red and green 'mists'). I mentioned the 'tennis' played between them – the rate of change in time of the red mist is determined by the curvatures of the green mist, and vice versa. This leads to the characteristic form of a momentum eigenstate, in which both mists have perfectly regular wave behaviour but with wave crests displaced relative to each other by a quarter of a wavelength. If the red crests are a quarter of a wavelength ahead of the green crests, the waves propagate in one direction and the momentum is in that direction. If the red crests are a quarter of a wavelength behind, the waves travel in the opposite direction and the momentum is reversed. We can call this *phase locking*. In a momentum eigenstate, there is perfect phase locking.

The semiclassical approach shows how two approximately phase-locked static waves can mimic evolution described by the time-dependent Schrödinger equation. In Figure 44 each of the two-dimensional wave patterns is nearly

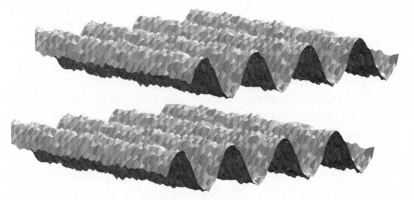

Figure 44 Two nearly sinusoidal wave patterns.

sinusoidal, and they are approximately phase-locked. These waves, being solutions of the stationary Schrödinger equation, are static – they do not move. But there is nothing to stop us (as in the example of the waves on the beach) from calling the direction along the axis perpendicular to the wave crests 'time' and the direction along the crests 'space'.

The key step now is to divide the total pattern of each wave into a regular part, corresponding to an imagined perfectly sinusoidal behaviour, and a remainder that is the difference between it and the actual (nearly sinusoidal) behaviour. Call this the *difference pattern* (there is one for each mist). If the condition of approximate phase locking holds, it turns out that the difference patterns satisfy with respect to our 'space' and 'time' an equation of the same form as the time-dependent Schrödinger equation, except for the appearance of one additional term. This term will have less and less importance, though, the more closely the assumptions of the semiclassical approach are satisfied.

In fact, the semiclassical approach offers the prospect of an explanation of time – in all its manifestations. It begins with a unified concept of things. Each point of Platonia is one distinct logically possible structure – it is one thing. The rules that make the structures make everything. Platonia is entire and eternal. No place in it is different from any other place, considered as something that is logically possible. But each structure is still a distinct individual. We see before us a true landscape whose every point is marked of necessity by individuality. It has striking topographic features. So there is a landscape, but nothing of a quite different nature that one might call time.

There is, though, one quite different element: a wave function. Schrödinger's enigmatic Ψ covers Platonia. Mist hovers over the eternal landscape. The static mist is a well-behaved solution – an eigenfunction – of the Wheeler–DeWitt equation. There is nothing here an unsuspecting bystander could say looks like time. You have seen mist on a landscape. Did it enter your head that such a thing could explain time? But it can, in principle. The static wave function, simply by its well-behaved response to the landscape it finds, may be induced into a regular wave-like pattern. If so, time can 'emerge' from timelessness. We shall see how the wave function enables the logically possible structures to interact – in a very real sense – with each other, thereby helping each other into an actual existence that seems to be deeply marked by time.

WHY DO WE THINK THE UNIVERSE IS EXPANDING?

This 'marking with time' brings us to the tricky part in the semiclassical approach. It is what led me to the notion of time capsules. This is a point at which my ideas part company from (comparative) orthodoxy. Two closely related difficulties convinced me that a radical step was needed. The first arises from a significant difference between the two Schrödinger equations. The complex time-dependent equation is actually two equations for two separate components – the red and the green mist. They play a kind of 'tennis' which tightly couples their behaviour and creates phase locking in any semiclassical solution. In contrast, the stationary equation is usually a real equation which does not couple the two components of the wave function.

The existence of two separate yet almost perfectly matched wave patterns is crucial in the semiclassical approach. The waves must be parallel, and the wave crests displaced by a quarter of a wavelength. In standard quantum mechanics this is a valid assumption. Indeed, it is imposed because the true primary equation is the time-dependent Schrödinger equation. The secondary stationary equation is just a short cut to tell us the distribution of the blue mist without having to find the red and green mists first. But they are there, and they are of necessity phase-locked.

But quantum cosmology gives us only the Wheeler–DeWitt equation. It is the primary equation, but as it stands it will give only a blue mist. We cannot assume some deeper equation hiding behind it that will give phase-locked red and green mists. The truth is that this part of the semiclassical approach assumes something that should be derived. Luckily, this difficulty threatens to undermine only that part of the semiclassical approach in which the specific structure of the time-dependent Schrödinger equation is recovered. The broad picture in which 'time' emerges from timelessness is not threatened. In fact, complex numbers, which appear in my account in the guise of the red and green mists, are so deeply ingrained in quantum mechanics that I feel fairly confident that this problem will be sorted out. What is needed is some independent argument which enforces the appearance of a complex wave function and a coupling between its components. That would then ensure the necessary phase locking.

Nevertheless, we must take care not to introduce inadvertently into quantum cosmology assumptions that may be valid only in ordinary quantum mechanics. This brings me to the second difficulty with the semiclassical approach. It concerns motion and our conviction that we experience it, and simultaneously the issue of where our sense of the passage of time comes from. To understand the answer to this question is to understand time. It is all very well for me to speak about static wave patterns in a mist that hangs over Platonia. Such patterns will indeed, where they are sufficiently regular, define unambiguously a direction that may be called 'time'. But even if the wave pattern is rather regular, we could not look at it and say that it distinguishes a direction of time. The one direction at right angles to the wave crests will look the same whichever way we face. There will be no signs set up on the distant horizons saying PAST and FUTURE. This is the issue we must now address.

It will be helpful to think about what it is that determines which way quantum wave packets move. Quantum mechanics is very different from classical mechanics in this respect. A classical initial condition consists of an initial position and an initial velocity. You know which way a particle will move because it is specified in the velocity. However, in quantum mechanics the initial condition is simply the values of the two components of the wave function, the red and green mists, everywhere at the initial time. Data like this seem to correspond to giving only the position in classical mechanics. Yet wave packets move under the rules Schrödinger prescribed.

In fact, the way in which a wave packet will move is coded in the relative positioning of the crests and troughs of the red and green mists. We see this most clearly in momentum eigenstates. If the red crests are ahead of the green crests, they go one way, but if the crest positioning is reversed, they go the other way.

As we have seen, states very like momentum eigenstates play a crucial role in the semiclassical approach. In all the original papers, these states also played another role – they were used to model situations corresponding to either expanding or contracting universes. All physicists and astronomers are convinced that we live in an expanding universe. There is certainly very good evidence in many different forms to support this view. The formalism of quantum cosmology must be capable of reflecting this aspect of the observed universe. There must be something that codes expansion or contraction of the universe or, rather (in the timeless interpretation), codes the observed evidence that leads us to say the universe is expanding.

All the models of Platonia that we have considered include a dimension that we may call the 'size' of the universe. In fact, instead of representing Triangle Land by means of the sides of triangles, we could equally well – and more appropriately here – use two angles (the third is found by subtracting their sum from 180°) and the area of the triangles. The area is one direction, or dimension, in Triangle Land. Expansion or contraction of the universe then corresponds to motion along the line of increasing or decreasing size. The size dimension begins at the point of zero size – what I have called Alpha, or the centre of Platonia – and then proceeds all the way to infinity.

In the semiclassical approach, it was rather reasonably assumed that the regular wave pattern needed for 'time' to emerge from timelessness would develop along the direction of increasing or decreasing size. This is a fair working hypothesis. What worried me was the way in which expanding and contracting universes were modelled – by analogy with momentum eigenstates in ordinary quantum mechanics. Expansion or contraction were supposed to be coded in the relative positions of wave crests.

It is certainly possible to imagine two static wave patterns – our red and green mists – whose crests are perpendicular to lines that seem to emanate from Alpha. This was done by nearly all the researchers who used the semiclassical approach, and they assumed that one relative positioning of the 'red' and 'green' crests would model a universe expanding out of the Big Bang, while the opposite positioning would model a universe headed for the Big Crunch (the name given to one possible fate of the universe, in which it recollapses to a state of infinite density and zero size). Thus, momentum-like semiclassical states were used to achieve three different things at once: the emergence of 'time', the recovery of the time-dependent Schrödinger equation, and modelling expanding and contracting universes. I believe that only the first is soundly based. I have some concern about the second. I think the third is definitely wrong.

The point is that the position of the 'green crests' ahead of or behind the 'red crests' by itself has no significance. In ordinary quantum mechanics the wave function depends not only on the spatial position but also on the time. What really moves wave packets is the relation of the time dependence to the space dependence. It is not the case that if in some wave packet the green crests are ahead of the red crests then the wave packet is bound to move one way. This happens only because the time-dependent Schrödinger equation is written in a particular form. But this is a pure convention. All observed phenomena are described just as well by an alternative choice, analogous to changing ends in tennis. The two choices are identical in their consequences. They only differ in the relative positions of the red and green crests, but this is offset by reversing the time dependence. The real physics is unchanged. Without the time dependence, the positions of the crests cannot determine the direction of motion.

But this presents us with a real dilemma in static quantum cosmology, in which there is no external time and no time dependence to determine which way wave packets move. There is simply no motion or change at all. We have to find a different explanation for why we think there is motion in the world and that the universe expands.

One thing is clear: the origin of our belief that the universe is expanding cannot be coded in the relative positioning of the crests of the two waves, for the designations 'red' and 'green' are purely conventional. The 'colours' could be swapped, and nothing observable would change. The argument that mere static positioning of crests can correspond to what we call expansion of the universe is a chimera. This was clearly recognized in 1986 by my German physicist friend Dieter Zeh, who commented that it has meaning only if an absolute time exists. It really is necessary to think very differently about these things if time is abolished once and for all as an independent element of reality.

THE IDEA OF TIME CAPSULES: THE KINGFISHER

From 1988 to 1991 I was absorbed by this issue. I became more and more convinced that a decisive new idea was needed, but for a long time could find no answer that satisfied me. I formulated the problem this way. I imagined myself watching some phenomena involving motion in a very essential and vital way – a display of acrobatics, say, or the flight of a kingfisher. I then imagined being struck dead instantaneously and my 'soul' being carried down to a kind of Plato's cave. Here I would find omniscient mathematicians examining a model of Platonia all covered with these red, green and blue quantum mists that I have asked you to conjure up in your mind's eye. They are examining the solution of the Wheeler–DeWitt equation corresponding to the universe in which I had just been taken from life. I then asked myself this: what precise thing in that mysterious pattern of mists blanketing Platonia corresponds to my being aware of seeing the kingfisher in flight? Where – in a timeless static world – is the appearance of motion coded? Where can I see the kingfisher's colours flashing in the sunlight?

As we have noted, in standard quantum mechanics the information about wave-packet motion is coded in the relative positioning of the red and green mists. This was the questionable assumption taken over in the semiclassical approach. However, there is much more to quantum mechanics than just the wave function at one instant (the pattern of red and green mists). We have already seen how time is needed if such relationships are to be translated into wave-packet motion. But even that is not enough, for the wave function acquires definite meaning only through prescriptions about the measurements that will be made on the system. These take the form of statements about the positions and construction of measuring instruments that behave classically and are external to the quantum system.

It is obvious that in quantum cosmology the whole superstructure of an external time, and of measuring instruments outside the considered system, must go. The instruments must be subsumed into the quantum system (which becomes the complete universe), and we must get to grips with a static wave function. Does this leave any scope for making a connection between actual experiences and the bare bones of embryonic quantum gravity as found by DeWitt?

I believe it does. Is not our most primitive experience always that we seem to find ourselves, in any instant, surrounded by objects in definite positions? Each experienced instant is thus of the nature of an observation, a discovery, even – we establish *where we are*. Moreover, what we observe is always a collection, or totality, of things. We see many things at once. In fact, most humans, indeed nearly all animals, have a wonderfully developed spatial awareness. In writing this book I have relied heavily on you possessing this gift – time and again I have asked you to imagine configurations of the universe as entities. They are all the places in Platonia.

When, therefore, I find myself in Plato's cave and see his demesne of Platonia laid out before me, I can, using my vivid memory of the kingfisher flashing between the banks of the stream where I stood, identify the instant in which death took me. By 'identify the instant', I mean recognize the configuration of riverbank, sunlight and shadow, rippled water and kingfisher's wings – all frozen in the position I last witnessed. As always, I insist that *instant of time* simply means configuration of the universe. This part of the problem of finding a connection between the psychical experience and the model of physical reality is relatively straightforward. There is little or no problem in the representation of position.

The real problem, then, is in the representation of motion. We seem to have exhausted all the resources of static quantum cosmology simply to put everything into place on the riverbank. Quantum mechanics does permit us to gain total information about position, but only at the expense of total loss of information about motion. We seem to have nothing left over to enable the kingfisher to fly. This is the crux of the matter. Classical physics presupposes both positions and motions, matching our experience that we see both at once. But quantum mechanics – in its present standard form – has this curious halving of the accessible data.

So how can we let the kingfisher fly? As few things delight me more than a kingfisher in flight, this is a matter of some interest to me. The answer that suddenly came to me in the summer of 1991 (which, of course, is a place in Platonia, not a time) was that the flight of the kingfisher is ultimately an illusion, though it rests on something that is very special and just as real as we take flight to be. It is flight without flight. Let me return to the imagery of the blue mist that shimmers over Platonia. It is easy to locate the instant of my death – I see the point in that great configuration space in which I stand on the bank of the stream. Now let me make an assumption in the hallowed tradition of Boltzmann: *only the probable is experienced.* The blue mist measures probability. Therefore, in accordance with the tradition, the blue mist must shine brightly at the point in Platonia in which I see the kingfisher frozen in flight above the water. I experienced the scene, so it must have a high probability. But there is still no motion.

I do not think there can be any. But there can be something else. As I mentioned in Part 1, nobody really knows what it is in our brains that corresponds to conscious experience. I make no pretence to any expertise here, but it is well known that much processing goes on in the brain and, employing normal temporal language, we can confidently assert that what we seem to experience in one instant is the product of the processing of data coming from a finite span of time.

This is all I need. It enables me to make the working conjecture that I outline in Part 1 – that when we think we see motion at some instant, the underlying reality is that our brain at that instant contains data corresponding to several different positions of the object perceived to be in motion. My brain contains, at any one instant, several 'snapshots' at once. The brain, through the way in which it presents data to consciousness, somehow 'plays the movie' for me.

Down in Plato's cave, thanks to the perfect representation of everything that is, I can look more closely at the point in the model of Platonia that contains me at the point of death. I can look into my brain and see the state of all its neurones. And what do I see? I see, coded in the neuronal patterns, six or seven snapshots of the kingfisher just as they occurred in the flight I thought I saw. This brain configuration, with its simultaneous coding of several snapshots, nevertheless belongs to just one point of Platonia. Near it are other points representing configurations in which the correct sequence of snapshots that give a kingfisher in flight is not present. Either some of the snapshots are not there, or they are jumbled up in the wrong order. There are infinitely many possiblilities, and they are all there. They must be, since there is a place in Platonia for everything that is logically possible.

Now, at all the corresponding points the blue mist will have a certain intensity, for in principle the laws of quantum mechanics allow the mist to seep into all the nooks and crannies of Platonia. Indeed, the first quantum commandment is that all possibilities must be explored. But the laws that mandate exploration also say that the blue mist will be very unevenly distributed. In some places it will be so faint as to be almost invisible, even with the acuity of vision we acquire in Plato's cave for things mathematical. There will also be points where it shines with the steely blue brilliance of Sirius – or the kingfisher's wings. And again my conjecture is this: the blue mist is concentrated and particularly intense at the precise point in Platonia in which my brain does contain those perfectly coordinated 'snapshots' of the kingfisher and I am conscious of seeing the bird in flight.

As I explained in Chapter 2, a time capsule, as I define it, is in itself perfectly static – it is, after all, one of Plato's forms. However, it is so highly structured that it creates the impression of motion. In the chapters that follow, we shall see if there is any hope that static quantum cosmology will concentrate the wave function of the universe on time capsules. As logical possibilities, they are certainly out there in Platonia. But will Ψ find them?

Latent Histories and Wave Packets

SMOOTH WAVES AND CHOPPY SEAS

All interpretations of quantum mechanics face two main issues. First, the theory implies the existence of far more 'furniture' in the world than we see. I have suggested that the 'missing furniture' is simply other instants of time that we cannot see because we experience only one at a time. The other issue is why our experiences suggest so strongly a macroscopic universe with a unique, almost classical history. In the very process of creating wave mechanics, Schrödinger found a most interesting connection between quantum and classical physics that cast a great deal of light on this problem. The interpretation he based on it was soon seen to be untenable, but it is full of possibilities and continues to play an important role. It is the starting point of other interpretations, including the one I advocate, so I should like to say something about it.

In the 1820s and 1830s, William Rowan Hamilton, whom we have already met, established a fascinating and beautiful connection between the two great paradigms of physical thought of his time – the wave theory of light and the Newtonian dynamics of particles. Cornelius Lanczos, a friend of Einstein and author of the fine book *The Variational Principles of Mechanics*, opens his chapter on these things with a quotation from Exodus: 'Put off thy shoes from off thy feet, for the place whereon thou standest is holy ground.' Let me quote Lanczos – he is not exaggerating:

> We have done considerable mountain climbing. Now we are in the rarefied atmosphere of theories of excessive beauty and we are nearing a high plateau on which geometry, optics, mechanics, and wave mechanics meet on

common ground. Only concentrated thinking, and a considerable amount of re-creation, will reveal the full beauty of our subject in which *the last word has not yet been spoken*. We start with … Hamilton's own investigations in the realm of geometrical optics and mechanics. The combination of these two approaches leads to de Broglie's and Schrödinger's great discoveries, and we come to the end of our journey.

The italics are mine. Lanczos's account does end with Schrödinger's discoveries, but I think it can be taken one step further. By the way, do not worry about the call for 'concentrated thinking'. If you have got this far, you will not fail now.

Hamilton made several separate discoveries, but the most fundamental result is simple and easy to visualize. Two characteristic situations are encountered in wave theory – 'choppy' waves, as on a squally sea, and regular wave patterns. Hamilton was studying the connection between Kepler's early theory of light rays and the more modern wave theory introduced by Young and Fresnel. Hamilton assumed that light passing through lenses took the form of very regular, almost plane waves of one frequency (Figure 45).

In optics, many phenomena can be explained by such waves. To do this, we need to know how the wave crests are bent and how the wave

Figure 45 An example of a regular wave pattern, showing wave crests and the lines that run at right angles to them. Such patterns are characterized by two independent quantities – the wavelength and the amplitude (the maximum height of the wave).

intensity, which is measured by the square of the wave amplitude (Figure 45), varies. In general, when the wave is not very regular, the ways in which the wave crests bend and the amplitude varies are interconnected, and it is not possible to separate their behaviour. However, as the behaviour gets more regular, the amplitude changes less and simultaneously ceases to affect the bending of the wave crests. Hamilton found the equation that governs the disposition of the wave crests in this case. Now known as the *eikonal equation*, it is the foundation of all optical instruments – microscopes, telescopes – and also electron microscopes. Indeed, numerous effects in optics are fully explained by the bending of the wave crests. However, other phenomena, above all the diffraction and spreading of light when it passes through a small orifice, can be explained only by the full wave theory. In these phenomena the regular pattern of wave crests is broken up.

We shall stick to phenomena in which the wave crests remain regular. Lines that run at right angles to such wave crests can be defined; they are easy to visualize (Figure 45). Hamilton's work in the 1820s showed that these lines correspond to the older idea of light rays, and that there are two seemingly quite different ways of explaining the behaviour of light and the functioning of optical instruments. In the older, more primitive way, light is composed of tiny particles (corpuscles) that travel along straight lines in empty space, but are bent in air, water and optical instruments (made of glass). The theory of light corpuscles works because the paths they take, along Kepler's light rays, coincide with the lines that run at right angles to the wave crests. This is the second of Hamilton's great discoveries: if light is a wave phenomenon, there are nevertheless many occasions in which it can be conceived as tiny particles that travel along these rays.

This insight led to the distinction between *wave optics* and *geometrical optics*, which uses light rays. Innumerable experiments show that only wave optics, in which light is described by waves, can explain certain phenomena. The earlier theory of light rays simply fails under these circumstances. Equally, there are many cases in which geometrical optics, with its Keplerian light rays, functions perfectly well. We see here the typical situation that arises when a new theory supplants an old one. The new theory invariably uses very different concepts – it 'inhabits a different world' – yet it can explain why the old theory worked as well as it did and why it is that it fails where it does. Where the wave pattern becomes irregular, geometrical optics ceases to be valid.

Geometrical optics shows how theories that explain many phenomena impressively and simply can still give a misleading picture. As my daughter learned on those frosty nights, this had happened in ancient astronomy. Ptolemy's epicycles gave a beautifully simple and successful theory of planetary motion, but were made redundant when Copernicus made the Earth mobile. Geometrical optics is another classic example of a 'right yet wrong' theory. In fact, with its confrontation and reconciliation of seemingly different worlds (particles and waves), it is one long, ongoing saga. It started with Kepler's optics, continued with the rival optical theories of Newton (particles) and Huygens, Euler, Young and Fresnel (wave theory), and reached a first peak with Hamilton. It burst into life again in 1905 with Einstein's notion of the light quantum, then went through another remarkable transformation in Schrödinger's 1926 discovery of wave mechanics. I believe this saga has not yet run its course, as I will explain in the next chapter.

Now we come to Hamilton's next discovery – the explanation of Fermat's principle of least time, the idea that did more than anything else to foster the development of the principle of least action.

HISTORY WITHOUT HISTORY

Figure 46 shows the wave crests of a light wave in a medium in which the speed of light is the same in all directions but varies from point to point, causing the wave crests to bend. The speed of light is less where the crests are closer together. Obviously, if some particle wanted to get from A to F in the least time, travelling always at the local speed of light, it would follow the curve ABCDEF. The individual segments of this curve are always perpendicular to the wave crests, and any deviation would result in a longer travel time. But this is also exactly the route a light ray would follow, cutting the wave crests at right angles. This was another great discovery that Hamilton made – that when geometrical optics holds, the wave theory of light can explain both Fermat's principle of least time and Kepler's light rays. The rays follow the lines of least travel time, and these are simultaneously the lines that always run perpendicular to the wave crests.

One of the most interesting things about geometrical optics is a connection it establishes with particles in Newtonian mechanics. The

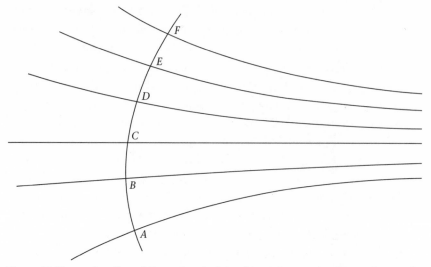

Figure 46 The explanation of Fermat's principle of least time under the conditions of a regular wave pattern, so that geometrical optics holds.

characteristic property of a moving particle is that it traces out a path through space. When regular wave patterns are present, wave theory creates similar one-dimensional tracks without any particles being present at all – the tracks of the light rays. Of course, in a strict wave theory the rays are not really 'there', but they are present as theoretical constructs. And many phenomena can be explained rather well by assuming that particles really are there. As John Wheeler would say, one has 'particles without particles', or even 'histories without histories'.

In fact, work that Hamilton did about ten years after his optical discoveries shows how apt such a 'Wheelerism' is. As we saw in Part 2, classical physics is the story of paths in configuration spaces. They are Newtonian histories. Hamilton thought about what would happen if for them only one value of the energy is allowed, and made a remarkable discovery. He found that just as light rays, which are paths, arise from the wave theory of light when there is a regular wave pattern, the paths of Newtonian dynamical systems can arise in a similar fashion. I need to spell this out.

Working entirely within the framework of Newtonian dynamics, Hamilton introduced something he called the *principal function*. All you need to know about this function is that it is like the mists on configuration space: at each point of the configuration space, it has a value (intensity), the variation of which is governed by a definite equation.

Hamilton showed that when, as can happen, the intensity forms a regular wave pattern, the family of paths that run at right angles to its crests are Newtonian histories which all have the same energy. They are not all the histories that have that energy, but they are a large family of them. Each regular wave pattern gives rise to a different family. Hamilton also found that the equation that governs the disposition of the wave crests, which in turn determine the Newtonian histories, has the same basic form as the analogous eikonal equation in optics. But whereas that equation operates in ordinary three-dimensional space, this new equation operates in a multidimensional configuration space.

Many physicists have wondered how the beautiful variational principles of classical physics arise. Hamilton's work suggests an explanation. If the principle that underlies the world is some kind of wave phenomenon, then, wherever the wave falls into a regular pattern, paths that look like classical dynamical histories will emerge naturally. For this reason, waves that exhibit regular behaviour are called *semiclassical*. This is because of the close connection between such wave patterns and classical Newtonian physics. It also explains the name of the programme discussed in the previous chapter.

All the things that this book has been about are now beginning to come together. A review of the essential points may help. We started with Newton's three-dimensional absolute space and the flow of absolute time. History is created by particles moving in that arena. Then we considered Platonia, a space with a huge number of dimensions, each point of which corresponds to one relative configuration of all the particles in the Newtonian arena. The great advantage of the concept of a configuration space, of which Platonia is an example, is that all possible histories can be imagined as paths. There are two ways of looking at the single Newtonian history that was believed to describe our universe. The first is as a spot of light that wanders along one path through Platonia as time flows. The spot is the image of a moving present. In the alternative view, there is neither time nor moving spot. There is simply the timeless path, which we can imagine highlighted by paint. Newtonian physics allows many paths. Why just one should be highlighted is a mystery. We have also seen that only those Newtonian paths with zero energy and angular momentum arise naturally in Platonia.

Hamilton's studies opened up a new way to think about such paths. It works if the energy has one fixed value, which may be zero, and introduces a kind of mist that covers the configuration space with, in general, variable intensity. In those regions in which the mist happens to fall into a pattern with regular wave crests, there automatically arise a whole family of paths which all look like Newtonian histories. They are the paths that run at right angles to the wave crests. If you were some god come on a visit to the configuration space and could see these wave crests laid out over its landscape, you could start at some point and follow the unique path through the point that the wave crests determine. You would find yourself walking along a Newtonian history. However, your starting point, and the path that goes with it, would have to be chosen arbitrarily, because precisely when the pattern of wave crests becomes regular, the wave intensity (determined by the square of the wave amplitude) becomes uniform. There would be nothing in the wave intensity to suggest that you should go to one point or another.

Hamilton's work opens up a way to reconcile contradictory pictures of the world. Quantum mechanics and the Wheeler–DeWitt equation suggest that reality is a static mist that covers Platonia. But all our personal experience and evidence we find throughout the universe speak to us with great insistence of the existence of a past – history – and a fleeting present. The paths that can be followed anywhere in Platonia where the mist does form a regular wave pattern can be seen as histories, present at least as latent possibilities.

I feel sure that the mystery of our deep sense and awareness of history can be unravelled from the timeless mists of Platonia through the latent histories that Hamilton showed can be there. But just how is the connection to be made? In the remainder of this chapter I shall explain Schrödinger's valiant, illuminating, but unsuccessful attempt to manufacture a unique history out of Hamilton's many latent histories. Then, in the next chapter, I shall consider the alternative – that all histories are present.

AIRY NOTHING AND A LOCAL HABITATION

When Schrödinger discovered wave mechanics he was well aware of Hamilton's work, since de Broglie had used the deep and curious connection between wave theory and particle mechanics in his own proposal. De Broglie's genius was to suggest that Hamilton's principal function was not just an auxiliary mathematical construct but a real physical wave field that actually guided a particle by forcing it to run perpendicular to the wave crests. Schrödinger sought to exploit Hamilton's work somewhat differently. His instinct was to interpret the wave function as some real physical thing – say, charge density. Of course, this could not be concentrated at a point, since its behaviour was governed by a wave equation, and waves are by nature spread out. Nevertheless, Schrödinger initially believed that his wave theory would permit relatively concentrated distributions to hold together indefinitely and move like a particle. His work led to the very fruitful notion of wave packets. These can be constructed using the most regular wave patterns of all – plane waves like the example in Figure 45. A plane wave has a direction of propagation and a definite wavelength. All the lines that run perpendicular to the wave crests are then latent, or potential, particle 'trajectories'.

Because the Schrödinger equation has the vital property of linearity mentioned earlier, we can always add two or more solutions and get another. In particular, we can add plane waves. Although each separate solution is a regular wave throughout space, when the solutions are added the interference between them can create surprising patterns. This makes possible the beautiful construction of Schrödinger's wave packets (Box 15).

BOX 15   Static Wave Packets

A wave with its latent classical histories perpendicular to the wave crests is shown at the top of Figure 47. Using the linearity, we add an identical wave with crests inclined by 5° to the original wave. The lower part of the computer-generated diagram shows the resulting probability density (blue mist). The superposition of the inclined waves has a dramatic effect. Ridges parallel to the bisector of the angle between them (i.e. nearly perpendicular to the original wave fields) appear, and start to 'highlight' the latent histories. In fact, these emergent ridges are the interference fringes that show up in the two-slit experiment (Box 11), in which two nearly plane waves are superimposed at a small angle, and also in Young's illustration of interference (Figure 22).

Much more dramatic things happen if we add many waves, especially if they all have a crest (are in phase) at the same point. At that point all the waves add constructively, and a 'spike' of probability density begins to form. At other points the waves sometimes add constructively, though to a lesser extent, and

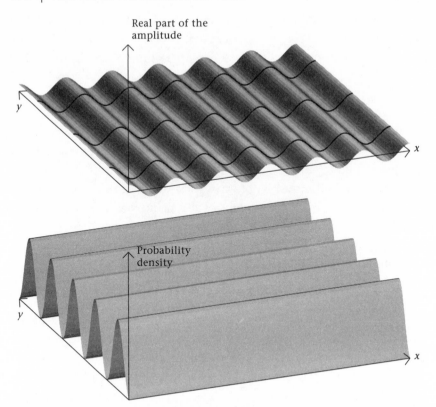

Real part of the
amplitude

Probability
density

Figure 47 If two inclined but otherwise identical plane waves like the one at the top are added, the figure at the bottom is obtained. The ridges run along the direction of the 'light rays' in the original plane waves. (The top figure shows the amplitude, the bottom the square of the added waves, since in quantum mechanics that measures the probability density.)

sometimes destructively. Wave patterns like those shown in Figure 48 are obtained.

Figure 48 brings to mind a passage in *A Midsummer Night's Dream* that has haunted poets for centuries:

And, as imagination bodies forth
The forms of things unknown, the poet's pen
Turns them to shapes, and gives to airy nothing
A local habitation, and a name.

The intersection of two wave fields does not result in any distinguished

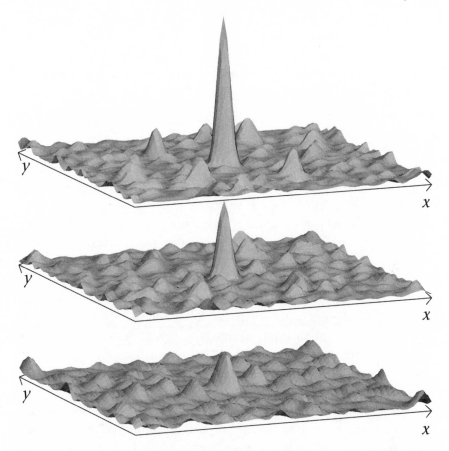

Figure 48 These wave patterns are obtained (from the bottom upward) by adding increasing numbers of plane waves oriented within a small range of directions. All waves have a crest where the 'spike' rises from the 'choppy' pattern. Their amplitudes also vary in a range, since otherwise 'ridges' like those in Figure 47 are obtained.

point, just a field of parallel ridges. There is no 'local habitation'. But if the crests of three or more waves intersect at a common point – so that the waves are in phase there – and their amplitudes are varied appropriately, then a point becomes distinguished. A localized 'blob' is formed. As Schrödinger realized with growing excitement in the winter of 1925/6, this begins to look like a particle.

The pièce de résistance is finally achieved if the waves of different wavelengths move and do so with different speeds. This often happens in nature. In most media – above all in vacuum – light waves all propagate with the same speed. However, in some media the waves of different wavelengths travel at different speeds. Since waves of different wavelengths have different colours, this can give rise to beautiful chromatic effects. In quantum mechanics, the waves associated with ordinary matter particles like electrons, protons and neutrons always propagate at different speeds, depending on their wavelengths. The relationship between the wavelength and the speed of propagation is called their *dispersion relation*.

Figure 49 has been constructed using such a dispersion relation. The initial 'spike' (wave packet) at the bottom is the superposition of waves of different angles in a small range of wavelengths. The dispersion relation makes each wave in the superposition move at a different speed. At the initial time, the waves are all in phase at the position of the 'spike', but the position at which all the waves are in phase moves as the waves move. The 'spike' moves! Its positions are shown at three times (earliest at the bottom, last at the top). This wave packet disperses quite rapidly because relatively few waves have been used in its construction. In theoretical quantum mechanics, one often constructs so-called Gaussian wave packets, which contain infinitely many waves all perfectly matched to produce a concentrated wave packet. These persist for longer.

It is a remarkable fact about waves in general and quantum mechanics in particular that the wave packet moves with a definite speed, which is known as the *group velocity* and is determined by the dispersion law. It is quite different from the velocities of any of the individual waves that form the packet. Only when there is no dispersion and all the waves travel at the same speed is the velocity of the packet the same as the speed of propagation of the waves. These remarkable purely mathematical facts about superposition of waves were well known to Schrödinger at the time he made his great discoveries – one of which was that this beautiful mathematics seemed to be manifested in nature.

SCHRÖDINGER'S HEROIC FAILURE

This led him to propose the *wave-packet interpretation* of quantum mechanics. His main concern was to show how a theory based on waves could nevertheless create particle-like formations. A potential strength of his proposal was that particle-like behaviour could be expected only above a certain scale. Over short enough distances, within atoms or in

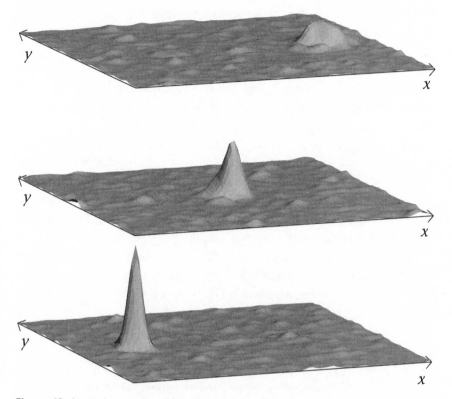

Figure 49 A moving wave packet obtained by adding plane waves having slightly different orientations, wavelengths and propagation velocities. The initially sharply peaked packet disperses quite quickly, as shown in the two upper figures.

colliding wave packets, the full wave theory would have to be used, but in many circumstances it seemed that particles should be present. With total clarity, which shines through his marvellous second paper on wave mechanics, he saw that if particles are associated with waves, then in atomic physics we must expect an exact parallel with geometrical optics. There will be many circumstances in which ordinary Newtonian particles seem to be present, but in the interior of atoms, for example, where the potential changes rapidly, we shall have to use the full wave theory. Schrödinger's second paper contains wonderful insights.

Unfortunately, his idea soon ran into difficulties. He had been aware of one from the start. For a single particle, the configuration space is ordinary space, and the idea that the wave function represents charge density makes sense. But he was well aware that his wave function was really defined for a system of particles and therefore had a different value for each configuration of them. I highlighted this earlier by imagining 'wave-function meters' in a room which showed the effect of moving individual atoms in models of molecules. It is difficult to see how the wave function can be associated with the charge density of a single particle in space.

Another problem actually killed the proposal. Although wave packets do travel as if they were a particle, they do spread. This was the one effect that Schrödinger failed to grasp initially. He actually did detailed and beautiful calculations for one special case – the two-dimensional harmonic oscillator, or conical pendulum. If a lead bob suspended on a weightless thread is pulled to one side and released, it will swing backwards and forwards like an ordinary pendulum. However, if it is given a sidewise jolt as well it will trace out an ellipse.

Schrödinger was able to show that for the quantum states corresponding to large ellipses it is possible to form wave packets that do not spread at all – the wave packets track round the ellipse for ever. This was a truly lovely piece of work, but misleading. Murphy's law tripped up Schrödinger. The harmonic oscillator is exceptional and is essentially the only system for which wave packets hold together indefinitely. In all other cases they spread, doing so rapidly for atomic particles. This doomed the idea of explaining particle-like behaviour by the persistence of wave packets, as Heisenberg noted with some satisfaction. (Most of the founding fathers of quantum mechanics defended their own particular directions with great fervour. Schrödinger hated quantum jumps, and found the extreme abstraction of Heisenberg's matrix mechanics 'positively repulsive'.)

However, the notion of a wave packet is beautiful and transparent, and has been widely and effectively used. This has tended to make people think that Schrödinger's original idea was still to a large degree right, and does explain why classical particle-like behaviour (restricted in its accuracy only by the Heisenberg uncertainty relation) is so often observed, especially in macroscopic bodies. There is one great difficulty, though. We can construct wave packets with strongly expressed particle-like properties, but we have to superimpose many different semiclassical solutions in just the right way. There must be a relatively small range of directions and wavelengths, adjustment of the wave amplitudes and, above all, coincidence of the phases of all waves at one point. Nothing in the formalism of quantum mechanics explains how this miraculous pre-established harmony should occur in nature. A single semiclassical solution might well arise spontaneously and naturally. But that will be associated with a whole family of classical trajectories, which exist only as formal constructions – they are at best latent histories. Quantum mechanics generally gives a wave function spread out in a uniform regular manner. Even if by some miracle we could 'manufacture' some wave packet, it would inevitably spread. Some further decisive idea is needed to explain how a universe described by quantum mechanics appears so classical and unique.

CHAPTER 20

The Creation of Records

HISTORY AND RECORDS

In Newtonian physics the notion of history is clear cut. It is a path, a unique sequence of states, through a configuration space. This picture is undermined in relativity and severely threatened in quantum mechanics, since the wave function in principle covers the complete configuration space. Almost all interpretations of quantum mechanics seek to recover a notion of history by creating or identifying in some way paths through the configuration space which are then candidates for the unique history that we seem to experience. This is a difficult and delicate exercise, since such paths simply do not belong to the basic quantum concepts. The methods used are quite varied, but they come in four main categories: the basic equations of quantum mechanics are modified (by ad hoc collapse of the wave function in the Copenhagen interpretation and by spontaneous physical collapse in some other interpretations); the equations are not changed but very special solutions are constructed (as Schrödinger attempted); extra elements are added to the quantum formalism (in so-called hidden-variable theories); or the equations and their solutions are accepted in full but it is asserted that the solutions in reality represent many parallel histories (Everett's many-worlds interpretation). None of these approaches is free of severe problems, some of which I have mentioned.

I suspect that the main difficulties arise because an important aspect of history has been ignored. Even if history is a unique succession of instants, modelled by a path in configuration space, it can be studied only through records, since historians are not present in the past. This aspect of history is not captured at all by a path. All the solutions of a Newtonian system correspond to unique paths, but they very seldom resemble the one history we do experience, in which records of earlier instants are contained in the present instant. This simply does not happen in general in Newtonian physics, which has no inbuilt mechanism to ensure that records are created. It is a story of innumerable histories but virtually no records of them. (I discussed this at the end of Chapter 1.)

In thinking about history, I believe we should reverse the priorities. Up to now the priority has been to achieve successions of states and to assume that records will somehow form. But nothing in the mechanisms that create successions ensures that records of them will be created. Now a record is a configuration with a special structure. Quantum mechanics, by its very construction, makes statements about configurations: some are more probable than others. This is especially apparent in the quantum mechanics of the stationary states of atoms and molecules. It determines their characteristic structures. In contrast, there is no way that quantum mechanics can be naturally made to make statements about histories. It is just not that kind of theory.

It is also interesting that classical physics makes only one crude distinction. Either a history is possible because it satisfies the relevant laws, or it is impossible because it does not. The possible continuous curves in the configuration space are divided into a tiny fraction that are allowed and the hugely preponderant fraction that are not. It is yes or no. Quantum mechanics is much more refined: all configurations are allowed, but some are more probable than others. By its very nature, quantum mechanics selects special configurations – those that are the most probable. This opens up the possibility that records, which are special configurations by virtue of their structure, are somehow selected by quantum mechanics. This is the possibility I want to explore in this and the following chapter. The aim is to show that quantum mechanics could create a powerful impression of history by direct selection of special configurations that happen to be time capsules and therefore appear to be records of history. There will be a sense in which the history is there, but the time capsule, which appears to be its record, will be the more fundamental concept.

THE CREATION OF RECORDS: FIRST MECHANISM

In the same conference in Oxford in 1980 at which Karel Kuchař spoke about time in quantum gravity, John Bell gave a talk entitled 'Quantum mechanics for cosmologists'. Among other things, he considered how records arise. This led him to describe a cosmological interpretation of quantum mechanics in which there are records of histories but no actual histories. Perhaps not surprisingly he rejected this as too implausible, but his account of how records arise is most illuminating. I shall reproduce it here in somewhat different terms, and then use it to propose an interpretation that is quite close though not identical to his, since Bell still assumed that the wave function of the universe would evolve with time. If this assumption is removed, as I believe it must be, Bell's interpretation becomes less implausible.

Bell illustrated how records are created in quantum mechanics by showing how elementary particles make tracks in detection devices. The essential principles had already been published, by Nevill Mott in 1929 and Heisenberg in 1930. As far as I am concerned, their work is more or less the interpretation of quantum mechanics, but surprisingly few people know about it.

It was stimulated by the Russian physicist George Gamow's theory of radioactive decay, put forward in 1928, in which alpha particles escape from radium nuclei by a process called tunnelling. The only detail we need to know is that Gamow represented an escaping alpha particle by means of an expanding, spherical wave function surrounding a radium nucleus. In accordance with the standard quantum interpretation, there is then a uniform density of the probability for finding the alpha particle all round the nucleus. In my pictorial analogy, blue mist spreads uniformly from the nucleus.

In those days, alpha particles were observed in devices called Wilson cloud chambers through their interaction with atoms, which they ionize by dislodging electrons, leaving the previously neutral atoms positively charged. The alpha particles invariably ionize atoms that lie more or less along a straight line emanating from the radioactive source. The excess positive charge of the ionized atoms stimulates vapour condensation around them, making the tracks visible. If we take Gamow's theory literally, there is something deeply mysterious about these tracks. If there really is a blue probability mist spreading out spherically all round the radium atom, why are atoms not ionized at random all over the chamber, wherever the blue mist permeates? How come they are ionized only along one line?

Standard quantum mechanics gives two answers, one much cruder than the other. In the crude answer (which is nevertheless very interesting, so I shall take a few pages to discuss it), only the alpha particle is treated in quantum-mechanical terms: the atoms of the cloud chamber are treated as classical external measuring instruments. They are used to 'measure the position' of the alpha particle, this being done by the ionization of an atom at some position. In accordance with the standard rules, any position measurement yields a unique position, after which the wave function will be concentrated at that position. The rest of the wave function will be instantaneously destroyed.

Now, atoms actually have a finite diameter, of about 10^{-8} centimetres. So the ionization of an atom is not a perfect position measurement, and this has important consequences for the alpha-particle tracks. It is helpful to think in terms of the blue mist. Before the measuring ionization happens, the blue mist is expanding outwards uniformly in all directions. When the first ionization occurs, it is as if a spherical shell has suddenly been placed round the atom. At one point on the shell there is a small hole through which the wave function can pass. This is the point at which the ionized atom is situated. It is only here that the wave function is not totally destroyed and can continue streaming on outwards. In fact, it does so in the form of a jet, which can be very narrow and accurately directed, especially if the alpha particle has a high energy.

At this point it is worth saying something about the diffraction of light. If monochromatic light (light of one wavelength) encounters an opaque screen with an opening, the result depends on its size. If the opening is large compared with the wavelength of the light, the screen cuts off all the light except at the opening and a more or less perfect 'pencil' of light – a beam – passes through. The width of the luminous pencil is equal to the width of the opening. However, if the opening is made smaller, diffraction comes into play and the beam of light spreads out, becoming very diffuse for a tiny opening. Diffraction effects are more pronounced for red light, with its longer wavelength, than violet light. Like light, alpha particles have an associated wavelength, which is very short for the ones produced in radioactive decay. Although ionization of the atom creates effectively a very small 'opening', the 'jet of wave function' that survives the wave-function collapse is narrow and concentrated in a cone with a very small opening angle (much less than a degree). The wave-function jet continues through the cloud chamber like a searchlight beam.

To simplify things, imagine that the cloud-chamber atoms are concentrated on uniformly spaced, spherical concentric shells surrounding the radium atom. The first ionization (quantum measurement and collapse) happens when the alpha particle's spherical waves reach the first shell. On the second shell, the alpha particle can ionize atoms only where its wave function has non-vanishing value. The atoms that can be ionized are located in the small spot that is 'lit up' by the 'beam' and hence lie rather accurately on the line joining the radium atom to the 'opening' in the first shell. The spot still contains many hundreds or thousands of atoms, any one of which can now be ionized. A second position 'measurement' of the alpha particle is about to be made.

The quantum measurement laws now tell us that one and only one of the atoms will be ionized. It is selected by pure chance – it can be anywhere in the spot. Once again, the entire wave function that 'bathes' the other atoms is instantly destroyed, and a new narrow beam continues outward from the second ionized atom. The same process of ionization, collapse and 'jet formation' is repeated at each successive shell. For an alpha particle with sufficient energy, this may happen hundreds or even thousands of times. A track is formed. It has some important features.

First, although it is nearly straight, there are small deflections at nearly all ionizations. It should not be supposed that the deflection occurs where the kink in the track suggests it did. This subtlety is illustrated in Figure 50. At each ionization and collapse a new cone of the wave function is created. It is not until the next ionization occurs that any actual deflection angle is selected. Until then, the complete cone of deflection angles is potentially present. As Heisenberg put it in a famous remark, the track is created solely by the fact that we observe the particle.

Second, quantum mechanics makes no predictions about the individual deflection angles. It merely predicts their statistical distribution, according to a law found by Max Born a few months after Schrödinger had created wave mechanics. Its form is determined by the structure of the atoms on which the scattering (deflection) of the alpha particle occurs. It is normally verified by making experiments with many different

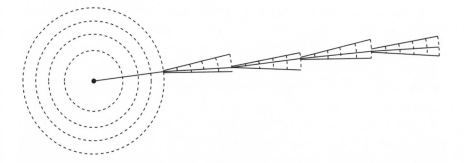

Figure 50 The creation of an alpha-particle track by successive ionizations. After each ionization a wave-function beam spreads out, but it is not until the next ionization occurs that the 'kink' is created.

alpha particles, the statistical distribution being built up by the repetition of many experiments over time. However, in principle it is possible to test the statistical predictions on a single track, especially if it contains thousands of ionizations.

Third, at each ionization the alpha particle loses a fraction of its energy, typically about one part in ten thousand. Since the energy is related to the particle's wavelength, it becomes progressively longer along the track. Just as diffraction effects are more pronounced for red than for violet light, this means that the deflection angles get progressively larger along the track. The nature of the track changes along its length – it starts to show quite large zigzags.

Bell comments on this first account of track formation that it 'may seem very crude. Yet in an important sense it is an accurate model of all applications of quantum mechanics.' Before we consider the second – infinitely more illuminating – account, we need to draw some conclusions and start to develop new ways of thinking about things, above all history.

THE PREREQUISITES OF HISTORY

The central question of this fifth part of the book is this: whence history?

What light does Bell's first account cast on this question? What are the essential elements that go into the creation of history? Bell's analysis promises to give us real answers to these questions, since an alpha-particle track can truly be seen as prototypical history. All the elements are there – a unique succession of events, a coherent story and qualitative change as it progresses. It even models birth – when the particle escapes from the radium atom – and demise – when it finally comes to rest. It literally staggers to its death. The laws that govern the unfolding of history are beautifully transparent. They combine, in an intriguing way, causal development – the forward thrust of the track – with unpredictable twists and turns governed only by probability. History is created by what looks like a curious mixture of classical and quantum mechanics – the continuous track and the twists and turns, respectively.

Three distinct factors together create history in this first account. First, the alpha particle emerges from the radium atom in a state that matches geometrical optics. Its wave function propagates outward in perfectly spherical waves of an extremely regular shape and with a very high frequency and short wavelength. This is a perfect example of a semiclassical solution. Hamilton's 'light rays' are the tracks that run radially outward from the radium atom, always perpendicular to the wave-function crests. Each of these tracks is a good simplified model of the one solitary track that eventually emerges.

I mentioned the ongoing saga of geometrical optics. Schrödinger attempted to create history by superimposing many slightly different semiclassical solutions in a wave packet that mimicked particle motion. We can now see that this attempt was doomed to failure, mainly because it attempted to create particle tracks using the quantum-mechanical properties of just one particle in isolation. The interaction of the particle with the environment played no role in Schrödinger's attempt, but is crucial in the account just given. We cannot begin to think of a track being formed without the atoms waiting to be ionized. Geometrical optics still plays a vital role because the very special semiclassical state ensures that sharply defined beams are created by the process of ionization and collapse.

We no longer need many semiclassical solutions: one semiclassical solution is now sufficient to create one history. Nevertheless, at least one semiclassical solution remains – and will remain – the prerequisite for history. The core mathematical fact discovered by Hamilton keeps reappearing and being used in different ways. I feel sure that this is the true deep origin of history – we have already seen alpha-particle tracks form before our eyes. Watch a little longer, and even Henry VIII and his six wives will appear.

The second element in Bell's account is collapse: crude, but effective. Little more needs to be said except that it is hard to believe that nature can behave so oddly. However, Bell's down-to-earth account does show up the artificiality of the quantum measurement rules. These are formulated for individual observables, and insist that measurement invariably results in the finding of a single eigenvalue of a chosen observable. But in the case of the alpha particle ionizing an atom, no pure measurement results – there is simultaneous measurement of both position and momentum (both with imperfect accuracy, so that the uncertainty relation is not violated).

The third element in the creation of history is low entropy: the initial state of the system is highly special. The alpha particle, which could be anywhere, is inside the radioactive nucleus; the countless billions of cloud-chamber atoms, which could be in innumerable different excited states, are all in their ground states. The only reason we are not amazed by such order is our familiarity with the special. What we have known from childhood ceases to surprise us. But even the experiencing of coherent thoughts is most improbable. Among all possible worlds, the dull, disordered, incoherent states are overwhelmingly preponderant, while the ordered states form a miniscule fraction. But such states, sheer implausibility, must be presupposed if history is to be made manifest – at least it is in the normal view of things.

The initial ordered state creates history and a stable canvas on which it can be painted. The special position of the alpha particle gives rise to its semiclassical state. The thousand or so atoms it ionizes stand out as a vivid track on the un-ionized billions. Photographed before dispersal, the track becomes a record of history. If a large proportion of the atoms were already ionized, such a track could hardly form, let alone stand out. We might claim that history had unfolded, but there would be no evidence of it.

Records are all we have. We have seen one account of their creation. Except for quantum collapse, it does not seem outlandish. But Bell gives a second, fully quantum account in which the monstrously multidimensional configuration space of the cloud chamber is vital. This story of history is amazing. The next section prepares for it.

THE IMPROBABILITY OF HISTORY

The cloud chamber is treated schematically as a collection of hydrogen atoms, each consisting of a nucleus – a single proton – and an electron. We ignore the fact (here not an issue) that all protons are identical, and so are all electrons. It is also reasonable to assume that the protons are at fixed points, and to treat only the electrons and the alpha particle quantum mechanically. The coordinates of each electron can be three mutually perpendicular distances from its proton. A real cloud chamber may contain 10^{27} atoms. It is daunting to contemplate a space with 3×10^{27} (+ 3 for the alpha particle) dimensions, but we must do our best if we are to get a true feeling for what is going on in quantum mechanics.

The really important thing here is that each configuration point represents one totality of all electron positions in the chamber. If we keep all the electrons fixed except one, which we move, it explores just three of the dimensions. In a much more modest way, there is an analogy here with our existence on the Earth: we live in three dimensions, but are normally restricted to the Earth's two-dimensional surface and do not normally move far in the third dimension. For the electron, the unexplored dimensions are not one but 3×10^{27}.

We can now think about representing an ionization track. The electron of a hydrogen atom has a characteristic probability distribution of diameter 10^{-8} centimetres around its proton. In quantum mechanics it is difficult to be certain about anything, but if we find a proton with no electron near it, this can indicate ionization – the electron has been torn away by the alpha particle. Imagine that we find a state of the chamber in which 1000 protons have no electrons near them; that these 1000 electron-less protons all lie more or less on a line between the decayed radium nucleus and the alpha particle; and that the statistics of the kinks along the line match Born's predictions for small-angle scattering. Naturally we should say that this is an alpha-particle track. It has all the appearances of recording quantum evolution with intermittent collapse. This state of the chamber, interpreted as an ionization track, is a perfect time capsule. Purely mathematically, it is a single point in a space. But the one point stands for a distribution of a huge number of electrons. As such, it is extraordinarily special – it is like a snapshot of history itself. If it could think, it would say, 'I am the track of an alpha particle moving in space and time through a cloud chamber.'

If the configuration space has innumerable dimensions, how much vaster is the number of its points. The overwhelming – hugely overwhelming – majority of the distributions they represent correspond to nothing interesting or striking. Sprinkled very thinly through this immense space are the distributions in which 1000 proton nuclei have no electrons near them. There are an incredible number of such distributions, but they are still much more thinly distributed than the stars in the sky. Within this already very thin company with 1000 ionizations are those for which the ionizations are all more or less on the line between the radium nucleus and its escaped alpha particle. But still these are not yet alpha-particle tracks. There is one more sieve – the scattering angles of the kinks must match Born's statistical distribution.

This piling of improbability upon improbability may seem pedantic, but I do want to bring home the sheer improbability of history. What immense creative power makes it? In addition, I am preparing the next step in the story of geometrical optics. For this, as I suggested earlier, it is helpful to start thinking of historical records as exceptional, specially structured points in configuration space: time capsules. Of course, if you look hard enough you can find not only them but all sorts of other things – pictures of Marilyn Monroe, more or less anything you like – but all such 'interesting pictures' are terribly thinly distributed. It is amazing that anything 'ferrets them out'. But causal quantum mechanics coupled with the incongruous collapse mechanism and a benign low-entropy environment can do the trick.

Before taking the next step, jettisoning collapse, we can add some refinements. In the collapse picture, we can not only mark (with 'paint') the configuration point that is the time capsule of the complete track. We can imagine a snapshot taken when only, say, 557 atoms have been ionized. The configuration point captured by it will also be a time capsule, and we can mark it too. If we mark in this manner all the stages – from no ionizations to all ionizations – all the corresponding time capsules will be different points in the configuration space. That is because they tell different stories, some of which only reach, say, the track's 'adolescence' or 'middle age'. Different configuration points necessarily represent different stories. However, they are joined up more or less continuously in a path, which represents an unfolding process.

If, like the god I imagined come to look at Platonia and its mists, we could 'see' the configuration space and the wave function sweeping over it, then in Bell's 'crude' account we should see a patch of wave function jigging its way along a track. The points along it are the complete cloud-chamber configurations with successively more ionizations. This configuration track is quite unlike the track that represents a history in Newtonian dynamics. For a single alpha particle, that is a track in three-dimensional space and the points along it, defined by three numbers, cannot possibly record history. In contrast, each of the points traced out in the big configuration space looks like a history of the three-dimensional track up to some point along it. An analogy may help. Doting parents take daily snapshots of their child and stick them day by day into a progress book. The progress book after each successive day is like each successive point along the track in the big configuration space: it is the complete history of the child up to that date. Similarly, a point along the track does not show the alpha particle at an instant of time, but its history up to that time.

THE CREATION OF RECORDS: SECOND MECHANISM

If experiments as in Bell's first account are repeated many times, a similar but different track will be photographed each time. Because quantum mechanics deals in probabilities, some tracks may well be more probable than others. Now imagine recording an alpha-particle track by 'marking' the corresponding configuration point with 'paint'. All configuration points that have been 'illuminated' in any of the experimental runs will be touched with paint, some many times. Because the instant of radio-active decay cannot be predicted, photographs taken at random will catch tracks of all 'ages' – birth, adolescence, middle age, old age. Eventually, many different points will have been touched by paint. A rich structure will have been highlighted. Perhaps the best way to picture this is as innumerable filaments, all emanating from the small region in the configuration space that represents the alpha particle trapped in the radium nucleus while all the cloud-chamber atoms are in their ground states.

It would be quite wrong to suppose that these filaments are so numerous that they fill the configuration space. That comes from confusion with ordinary three-dimensional space. It is always dangerous to take analogies too literally, but if we are going to try to use images, it is better to think of the structure that is formed in the configuration space by the points that have been 'touched with paint' as being more like strands of a spider's web spun out in the reaches of interstellar space with huge gaps between them. Such a structure is then a record of innumerable experiments interpreted in the first 'crude' way.

One more comment. So far, we have considered only single tracks. But in modern experiments a single particle colliding with a detector particle can create many secondary particles. These also make tracks simultaneously in the detector. A single quantum event gives rise to many tracks. If a magnetic field is applied the tracks are curved by different amounts depending on the particle masses, charges and energies. Beautiful patterns, representing quite complicated histories, are created (Figure 51). This multitrack process in ordinary space is still represented by one track in configuration space. History, no matter how complicated, is always represented by a single configuration path; records of that history, which may be very detailed and more or less pictorial (actual snapshots), can readily be represented by a single configuration point. A library containing all the histories of the world ever written is just one point in the appropriate configuration space.

We now come to the more sophisticated account of alpha-particle interaction with a cloud chamber. The entire process is treated quantum

Figure 51 Multiple tracks of elementary particles created by a single quantum event. The swirls and curved tracks arise from the effect of a magnetic field on the charges of the particles created.

mechanically – as wave-function evolution in a space of around 10^{27} dimensions. Initially, before the alpha particle escapes, the wave function (of all the electrons and the alpha particle) is restricted to a rather small configuration region. In the crude collapse picture, alpha-particle escape and track formation is represented as a 'finger' of wave function that suddenly emerges from it and rushes through the configuration space like a rocket shooting through the sky.

In the new picture, with everything treated quantum mechanically and no collapse, an immense number of wave-function 'fingers' emerge almost at once and race in a multitude of directions across the configuration space. Each follows more or less one of the tracks of the scenario with collapse. All the tracks are traced out simultaneously. It is like one of those spectacular fireworks that explodes and shoots out a blazing shower in all directions. This is what we should observe if we could see the wave function bursting out from its original confines into the great open spaces of Platonia.

It is not easy to explain why it behaves like this, but let me try. The most important thing is that a configuration space is not some blank open space like Newton's absolute space, but a kind of landscape with a rich topography. Think of the wave function pouring forth like flood-water sweeping over a rocky terrain, whose features deflect the water. It will help if you look again at Triangle Land (Figures 3 and 4). It is bounded by sheets and ribs, and is the configuration space for just three particles. The configuration space for 10^{27} particles is immensely more complicated. Things like the ribs and sheets that appear as boundaries of Triangle Land occur as internal topography in Platonia, which is traversed by all kinds of structures. The rules that govern the evolution of the wave function force it to respond to this rich topography. The wave-function filaments are directed by salient features in the landscape.

Now that we have some idea of how the 'firework explodes', we can think about its interpretation. The problem is that we never see configuration space. That is a 'God's-eye' view denied to our senses – but fortunately not to our imaginations. We also never see a solitary alpha particle making many tracks at once: all we ever see is one track. How is this accounted for in the second scenario? By the same device as before – by collapse. In the first scenario, the alpha particle was in many different places in its configuration space simultaneously before we forced it to show itself in one region. This was done by making it interact with an atom. This, most mysteriously, triggered collapse, which was repeated again and again.

In the second scenario, the complete system is, after a time sufficient for the ionization of 1000 atoms, potentially present at many different places in its huge configuration space. The wave function is spread out over a very large area, though concentrated within it, in tiny regions. All the points within any of these regions is like a snapshot of an ionization track, all differing very slightly (and hence represented by different points within a small region). There is an exact parallel between the alpha particle in the first scenario being at many different places before the first collapse-inducing ionization and the state now envisaged for the complete system of cloud chamber and alpha particle. It too is in many different 'places' at once.

We can now collapse this much larger system by making a 'measurement' on it to see where it is. This is often done simply by taking a photograph of the chamber. It catches the chamber in just one of its many possible 'places'. And what do we find? A chamber configuration showing just one ionization track, corresponding to one of the points within one of the tiny regions on which the wave-function mist is concentrated. We have collapsed the wave function, but this time onto a complete track, not onto one position of one particle.

If such experiments are repeated many times, the tracks obtained are found to be essentially the same as the tracks in the first scenario. There are in principle small differences, which come about because the evolution is not quite the same in the two cases – in the latter case the tracks can interfere to some extent, but in general the final results are more or less the same despite the very different theoretical descriptions.

The reason for this is that seeds of the many different tracks – different histories – are already contained in the initial wave function. A concentrated wave function necessarily spreads, and if this happens in a large enough configuration space under low-entropy conditions it can excite many different configurations that embody records of many different histories. There is a snowball effect. We start with many small snowballs, the different possibilities for the alpha particle at the beginning of the process. Each possibility then becomes associated – entangled – with a different track. This is rather like many different snowballs picking up snow. Subject always to a pervasive quantum uncertainty, a fuzziness at the edges, these are Everett's many worlds. The distinctness of these different worlds, the different histories, is determined by the extent to which part of the system (the alpha particle in this case) is in the semiclassical (geometrical-optics) regime.

It is the near perfection of the initial semiclassical state of the alpha particle that creates such sharply defined histories and ensures that two such different scenarios give more or less the same results. This is ultimately the reason why the notorious *Heisenberg cut* – the position at which we suppose the quantum world to end and the external, non-quantum world of classical measuring instruments to begin – can be shifted in such a bewildering manner. As Bell remarks, for practical purposes it does not matter much where we place the cut to determine where collapse occurs, since the end results are much the same. In either case, the appearance of history is created by interaction between the semiclassical part and the remaining, fully quantum system. The resulting correlation forces the quantum system into a very special state.

It is really almost miraculous how the classical histories, latent as very abstract entities within a semiclassical state of the alpha particle when it is considered in isolation, force the wave function of the remainder of the system (the cloud chamber) to seek out with extraordinary precision tiny regions of its vast configuration space. When these regions – or, rather, the points within them – are examined, they turn out to represent configurations that are snapshots of tracks. They are records of histories.

So this is the next twist in the saga. First Hamilton found families of classical, particle-like histories as 'light rays' in a regular (semiclassical) wave field. Then Schrödinger tried to mimic particle tracks by superposing many slightly different semiclassical solutions to create just one wave packet – the model of a single particle. It was rather hard and contrived work for a meagre – but still very beautiful – result. However, it immediately slipped through his fingers. But then Heisenberg and Mott showed that quantum mechanics could work far more effectively as the creator of history than Schrödinger had ever dreamed. Now one single semiclassical solution generates (before the final collapse) many histories. Instead of Schrödinger's contrived

Many semiclassical solutions → One history

we have natural organic growth:

One semiclassical solution → Many records of histories

CHAPTER 21

The Many-Instants Interpretation

MANY HISTORIES IN ONE UNIVERSE

The story goes on. We have put only the cloud chamber into the quantum mill – can we put the universe, ourselves included, in too? That will require us to contemplate the ultimate configuration space, the universe's.

You can surely see where this is leading. Now the snowballs can grow to include us and our conscious minds, each in different incarnations. They must be different, because they see different tracks; that makes them different. These similar incarnations seeing different things necessarily belong to different points in the universal configuration space. The pyrotechnics of wave-function explosion out of a small region of Platonia – the decay of one radioactive nucleus – has sprinkled fiery droplets of wave function at precise locations all over the landscape. (What an awful mixing of metaphors – snowballs and sparks! But perhaps they may be allowed to survive editing. The snowballs are in the configuration space, the sparks in the wave function. This is a dualistic picture.)

And now to the great Everettian difference: collapse is no longer necessary. Nothing collapses at all. What we took to be collapse is more like waking up in the morning and finding that the sun is shining. But it could have been cloudy, or cloudy and raining, or clear and frosty, or blowing a howling gale, or even literally raining cats and dogs. When we lay down to sleep in bed – when we set up the alpha-particle experiment – we knew not what we should wake to. What we take to be wave-function collapse is merely finding that this ineffable self-sentient something that we call ourselves is in one point of the configuration space rather than another. When we observe the outcome of an experiment, we are not watching things unfold in three-dimensional space. Something quite different is happening. We are finding ourselves to be at one place in the universal configuration space rather than another. All observation, which is simultaneously the experiencing of one instant of time, is ultimately a (partial) locating of ourselves in Platonia. Each of our instants is a self-sentient part of a Platonic form.

The coherence of this picture hangs on the ability of the universal wave function to seek out time capsules in Platonia that tell a story of organic growth. All stories are in Platonia, some bizarre beyond the dreams of Hieronymus Bosch or modern surrealists. The history that we experience may have its horrors, but it is extraordinarily coherent and self-consistent. The first task of science is to save the appearances. So, first and foremost, we need to find a rational explanation for the habitual miraculous experiencing of time capsules, these freighters of history. This is where the probability density of the wave function, its shimmering blue mist, plays such a crucial role. Because apparent records of all histories – and a mind-numbing multitude of non-histories – are present in Platonia, we shall not have an explanation of the appearances worthy of the name unless the blue mist shines brightly over time capsules of the kind we know so well from direct experience. And it should not shine brightly anywhere else. We shall then have a theory that does truly save the appearances. Bell's analysis hints that universal quantum cosmology might be that theory.

It is time to take stock once more. First, we muster the interpretations of quantum mechanics. How do they look in the light of Bell's analysis? What appearances do they save and how well do they do it? There are two minimum requirements of an interpretation – it must explain why we see just one world (Einstein's Moon problem) and it must explain why we think it has a history. The latter is the harder task. However, it may be important not to ask for too much. To save the appearances, we do not have to create a unique history: we need only explain why there seems to be a unique history. That was Everett's insight. If we can stand back from our parochial prejudices, a theory which can achieve that is already little short of miraculous.

Except for many-worlds variants, all the interpretations strive for the severe criterion of only one history. They were created for that and all achieve it by brute force. History is created by repeated strangling of the wave function (Copenhagen and physical collapse) or by adding incongruous extras: the so-called hidden variables. The German *doppelt gemoppelt* means messing things up by doing them twice over. In their anxiety to recover a unique history, the proponents of these interpretations crudely impose one history on a theory that can already create many histories which are autonomous – and hence each unique in our experience – by a beautiful natural mechanism. Hamilton's discovery makes it inescapable that histories are latent in the quantum formalism. It is just a matter of coaxing them out into the open.

BELL'S 'MANY-WORLDS' INTERPRETATION

From his discussion of alpha-particle tracks, Bell turned to a remarkable cosmological interpretation of quantum mechanics. It makes essential use of the notion of time capsules and is therefore very similar to the interpretation I shall present in the final chapter. Bell saw it as a way of retaining Everett's idea that the wave function never collapses without proliferating worlds.

Bell claimed that the really novel element in Everett's theory had not been identified. This was 'a repudiation of the concept of the "past", which could be considered in the same liberating tradition as Einstein's repudiation of absolute simultaneity'. Obviously, something exciting is in prospect, and Bell does not disappoint. He looked for the quantum property that enabled Everett to make his many-worlds idea plausible, and pointed out that the accumulation of mutually consistent records is a vital part of it. This recognition had led Bell to his analysis of the formation of alpha-particle tracks, which have the obvious interpretation that they are records of alpha-particle motion. He showed that 'record formation' is a characteristic quantum property. At least under cloud-chamber conditions, the wave function concentrates itself at configuration points that can be called records. Although Bell did not use my term, such points are manifestly time capsules. He noted that Everett's interpretation could not even be formulated were it not for the wave function's propensity to find them.

He then attacked head-on the conventional notion of history inherited from classical physics as a continuous path through configuration space. This might make sense if, god-like, we could see all time and the configuration space with history highlighted as a path in it by a 'thread' or 'paint'. But our only access to the past is through records. As Bell says, 'We have no access to the past. We have only our "memories" and "records". But these memories and records are in fact *present* phenomena.' Our only evidence for the past is through present records. If we have them, the actual existence of the past is immaterial. It will make no difference to what we know. Hence 'there is no need whatever to link successive configurations of the world into a continuous trajectory'.

His 'Everettian' interpretation is this: time exists, and the universal wave function Ψ evolves in it without ever collapsing. Because Ψ has the propensity to seek out time capsules, it will generally be concentrated on them. Real events are actualized as follows. At each instant of time, Ψ associates a definite probability (the intensity of the blue mist in my analogy) with each configuration. At any instant, just one event is actualized at random in accordance with its relative probability. The higher the probability, the greater the chance of actualization. Since time capsules have the highest probabilities, they will generally be selected.

Sentient beings within them will possess memories and records that convince them they are the product of history. But this will be an illusion. In reality, the points realized at successive instants of time are chosen randomly and jump around in a wildly unpredictable manner in the configuration space. The sentient beings within the actualized points have memories of quite different histories. It is all very bizarre, though within each randomly selected time capsule the memories and records tell a most consistent story. Bell rejected his 'many-worlds' interpretation as too absurd:

> Everett's replacement of the past by memories is a radical solipsism – extending to the temporal dimension the replacement of everything outside my head by my impressions, of ordinary solipsism or positivism. Solipsism cannot be refuted. But if such a theory were taken seriously it would hardly be possible to take anything else seriously. So much for the social implications. It is always interesting to find that solipsists and positivists, when they have children, have life insurance.

This is all very entertaining – and I too have children and life insurance – but these are just the kind of *ad hominem* quips that were tossed at Copernicus and Galileo. I do believe that Bell came close to a viable cosmological interpretation of quantum mechanics, and should have kept faith with his title ('Quantum mechanics for cosmologists'). But he left the cosmologists with nothing. Later he gave warm support to one of the theories in which wave-function collapse is a real physical process. In it, the propensity of the quantum-mechanical wave function to find time capsules plays no role. History is created by a succession of actually realized states. It is there with or without any record of it.

From the way Bell wrote in 1980, either he was unaware of the Wheeler–DeWitt equation and the possibility that the universal wave function is static, or he dismissed this without mention. It would be interesting to know how he would have reacted to the idea – he seems to have had a somewhat Newtonian notion of time. Sadly, he died several years ago, so we cannot ask him. I regret this especially since his 1980 proposal is very close to mine in two of its three main elements. He may have believed in time, but his emphasis on memories and records and their rather natural occurrence in the quantum context are valuable support for me. So are his views on ontology and psychophysical parallelism. This is the third common element.

In discussing Everett's theory, I mentioned the so-called preferred-basis problem. This arises from transformation theory: a quantum state simultaneously encodes information about mutually exclusive properties. Viewed one way, it gives probabilities for particle positions; viewed another, it gives probabilities for their momenta. It is impossible to extract this information simultaneously and directly by, so to speak, 'looking at the system'. We must let the system interact with instruments. Depending on how the instruments are arranged, we can extract information about either the positions or the momenta, but not both at once. The ambiguity becomes especially acute if the instruments are treated quantum mechanically. We cannot say what state they are in or what they are measuring.

Bell advocated a simple and robust answer to this in many of his writings, including his 1980 paper: the complete system formed by the particles and the instruments measuring them is always defined in the last resort by positions. In any quantum state, different sets of positions are present simultaneously, but it is always positions that are present. The different kinds of quantum measurement, giving alternatively position-type and momentum-type outcomes, arise because the same sets of positions of the measured system are made to interact with characteristically different sets of instrument positions. Everything is ultimately inferred from positions. This is exactly my position. Platonia is the universal arena. To Bell's arguments – and gut conviction – for this standpoint I would add the impossibility of obtaining a satisfactory theory of inertia and time unless positions are fundamental.

Now, what did Bell regard as the physical counterpart of psychological experience? Is it in the wave function, as Everett and many others have assumed, or in matter configurations? Bell, like myself, opts for the latter: 'It is … from the **xs** [the configurations], rather than from ψ, that in this theory we suppose "observables" to be constructed. It is in terms of the **xs** that we would define a "psycho-physical parallelism" – if we were pressed to go so far.' Although Bell does not spell out his parallelism too explicitly – he does not seem to want to be 'pressed' too far – it is clear from the way he makes memories and records responsible for our idea of the past, rejecting any 'thread' connecting configurations at different times, that subjective awareness of both positions and motions of objects must be derived from the structure in one instantaneous configuration. The self-sentient configurations must be time capsules. Not only the kingfisher but also the appearance of its flight must be in one configuration, for nothing else would be logically consistent. The main lessons I draw from Bell's paper are incorporated in the many-instants interpretation that I favour.

THE MANY-INSTANTS INTERPRETATION

This is based on a conjecture that I shall try to justify in the next chapter. Here I simply assume it. It is that the universe is described by an equation of Wheeler–DeWitt type, which may have one or many solutions, and that each of its well-behaved solutions concentrates its probability density on time capsules. Bell showed that this does happen if time exists, and if evolution is real and commences from a low-entropy state. Since I deny time, I cannot appeal to a special initial state. There is only one state and no evolution. That is the problem for the next chapter; here I want to describe the kind of state I conjecture and how it must change our view of history.

Most important is a distinction between two different kinds of variable. Bell showed how the alpha-particle semiclassical state contains latent histories which then become entangled with the cloud-chamber electrons.

The electrons could be in a huge number of different configurations, but in the Mott–Heisenberg solution the only configurations with high probability are those that look like alpha-particle tracks. Something similar must happen in cosmology, but there is a difference.

Imagine a swarm of 5000 bees. Its configuration space has 15,000 dimensions. However, from a distance we cannot see the individual bees, only the overall position of the swarm and, say, its size (radius). These are four dimensions of the configuration space. In such situations a few of the configuration-space dimensions describe the system's large-scale properties, and the remaining, much more numerous dimensions describe the fine details. The corresponding large-scale and small-scale configuration spaces are illustrated in Figure 52.

Any point in Figure 52 represents a possible position of all the bees. Horizontal motion from a point changes the swarm's position and size without changing the relative position of the bees within it. Vertical displacement leaves the swarm's position and size unchanged but rearranges the bees. Since this can happen in so many ways, each vertical point actually represents multitudinous possibilities. Alas, we have only the vertical to represent them. Also, to make even a moderately realistic model of the universe, the horizontal positions should represent the positions of not just one swarm but many. Imagine, say, 100 swarms. Each horizontal position then represents one relative arrangement of their positions and sizes as complete units. Different vertical positions having the same horizontal position then correspond to all rearrangements of the bees that leave the swarms as they are. This is very schematic, but it is sufficient to explain the scheme.

If the wave function of the universe is static, quantum cosmology reduces to the question of how its values are distributed in Platonia. For the moment, I shall simply give you my guess; arguments for it come later. My guess is a special distribution closely similar to the cloud-chamber one described by Bell. In most of Platonia the wave function has extremely small values – the blue mist has negligible intensity. However, in a few special regions, distributed over a large area, the blue mist's intensity is, relatively, hugely higher. These regions correspond to some arrangement of the swarms, determined by the horizontal position, and to the detailed positions within them, determined by the vertical positions. It is in the probabilities for these detailed positions that the blue mist is extraordinarily selective. The probabilities for the horizontal positions are relatively uniform over quite large regions. By themselves, they represent a dull state of affairs. The situation is transformed by the configurations that specify the fine details within the swarms. At the very rare configurations where the blue mist shines brightly, the fine details look like records of a history of the swarms as complete units. They suggest that the swarms have moved in a classical history from some past up to a present instant, in the position they now occupy.

This is illustrated in Figure 52, in which the points X and Y in Platonia have large-scale positions B and D. The fine details at X and Y seem to represent records of how the swarms have moved from earlier configurations A and C along curves AB and CD in the large-scale (horizontal) configuration space. They seem to be records of these large-scale histories. The blue mist has a high intensity not only at X and Y but also, for example, at P and Q, at which points the fine details suggest they represent records up to the intermediate stages E and F in the histories AB and CD.

By no means all details need represent history. Footprints in the sand on a wide beach record the movements of people who have walked on it, but over much of the beach there need be no apparent records. Think again of the number of atoms in a pea. A tiny fraction of them can easily record the pea's history up to its current present. The huge numbers we confront in physics explain why we may have wrong ideas of what history actually is. We may have jumped to a conclusion too quickly.

In the Newtonian picture, in which history is a curve in configuration space, it is extremely hard to understand how records arise. Even if a single curve is realized, any point on it could have any number of histories passing through it. How can one instantaneous configuration of particles suggest the motions that they have? However, if we keep an open mind about the laws that determine things, a fraction of a pea's atoms may well seem to record a history of its large-scale features. This does not mean that all its atoms had a unique history. Without change in the pea's large-scale structure, the same large-scale history could be coded in innumerable different ways by only a tiny fraction of its atoms. In the imagery of Figure 52, there will be a whole cloud of points X in the configuration space that correspond to the same large-scale configuration and to the same history up to it. The different points in the cloud simply code the same history in different ways. What is more, for each point along the large-scale history AB there will be a corresponding cloud of points that record the same history up to that point in different ways.

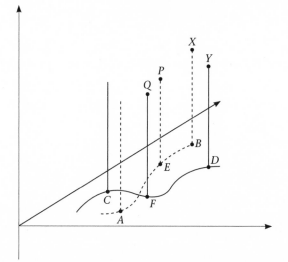

Figure 52 The division of Platonia. The horizontal dimensions represent the large-scale configuration space, and the single vertical dimension represents the small-scale space. The 'horizontal curves' (*AEB* and *CFD*) represent histories of the large-scale features. Each point like *A* represents, say, the overall position and size of a swarm of bees. In contrast, each of the points *Q*, *P*, *X*, *Y* on the vertical lines represents the huge number of small-scale details.

There will be a 'tube' of such points in the configuration space. No continuous 'thread' joins up these points in the tube into Newtonian histories. The points are more like sand grains that fill a glass tube. Each grain tells its story independently of its neighbours. In any section of the tube, the grains all tell essentially the same story but in different ways, though some may tell it with small variations.

I think this is the way to think about history in quantum stasis. Could we but see the picture – all Platonia with its misty crannies – we should see it as it is: the lawful definite world for which Einstein, like so many physicists, longed. But it is a timeless book full of different stories that tell of time. Quantum mechanics can create the appearance of multiple histories. However, will it in quantum cosmology? Its conditions are not quite Mott and Heisenberg's conditions.

CHAPTER 22

The Emergence of Time and its Arrow

CAUSALITY IN QUANTUM COSMOLOGY

John Bell's account of time-capsule selection contains a very large configuration space, time, the wave function and its equation (the time-dependent Schrödinger equation) and a special initial state. This last is most important. If quantum cosmology is static, something else must replace it. We cannot impose an initial condition in the past because there is no past. But we can try something similar. Suppose that the universal configuration space had only three dimensions and not the monstrous number I have so often asked you to consider. We could then specify the wave function on a two-dimensional plane in that three-dimensional space, and use the equation satisfied by the wave function to find it at other points. This is like evolving a state in time except that the evolution is in the third, spatial direction.

If we attempt this in ordinary quantum mechanics with the stationary Schrödinger equation, which in some respects at least is like the Wheeler–DeWitt equation, the wave function starts to misbehave sooner or later. Either it becomes infinite, or it cannot be evolved continuously, or some other disaster happens: it ceases to be 'well behaved'. The remarkable and exciting discovery that Schrödinger made was that the hydrogen atom does have a very special set of solutions that are well behaved everywhere and for which therefore no disaster happens. These very special states correspond exactly to the negative-energy states of the hydrogen atom. He had explained what had hitherto been one of the deepest mysteries of physics – the spectral lines of atoms and molecules.

My main interest here is the transformation of our notions about causality that a solution of this kind could represent. The traditional view is that what happens now was 'caused' by some state in the past. There is always arbitrariness in this picture because the past state is arbitrary. But suppose the world is described instead by a solution of some Wheeler–DeWitt equation that is everywhere well behaved in Schrödinger's sense. I have already pointed out that such solutions are ultra-sensitive to the domain on which they are defined – otherwise they could not remain well behaved everywhere. Such solutions present a kind of pre-established harmony.

The Wheeler–DeWitt equation then constitutes the rules of a game played in eternity. The wave function is the ball, Platonia is the pitch. If a well-behaved solution exists, then only two things can have conspired to create it: the rules of the game and the shape (the topography) of the pitch. In contrast, Bell's time capsules are created by the rules, time, the topography and a special initial condition. What a prize if we could create time capsules by the rules and the shape of the pitch alone! Arbitrary, vertical causality (through time) would then be replaced by timeless horizontal and rational causation – across Platonia.

SOCCER IN THE MATTERHORN

It is possible. There are plenty of time capsules in Platonia. It is not just time and the special initial conditions that enable the wave function to find time capsules. The rules of the game and, above all, the pitch size and topography are most conducive to it. Indeed, the configuration space is a prerequisite. As Nevill Mott remarked, 'The difficulty that we have in picturing how it is that a spherical wave can produce a straight track arises from our tendency to picture the wave as existing in ordinary three-dimensional space, whereas we are really dealing with wave functions in the multispace formed by the co-ordinates both of the alpha-particle and of every atom in the Wilson chamber.' What interests me now is not so much the dimensions as the pitch's shape. What follows is speculation. Mine. I am not aware that anyone else has made it (though Dieter Zeh considered something rather similar). I have lectured several times on the idea, and in 1994 published quite a long paper on it in the journal *Classical and Quantum Gravity*. A problem with the idea is that as yet it is purely qualitative. Physicists rightly want to see real calculations (which, alas, are bound to be difficult), not mere speculation, before they endorse an idea. But the more I think about the idea, the more plausible, indeed almost inescapable, it appears to be. It is about the origin of the arrow of time – and time itself.

The arrow of time, manifested in the ubiquity of time capsules, is a colossal asymmetry. It is well-nigh inexplicable in time-symmetric physics, the present rules. Since Boltzmann's age, it has towered there, an unscaled Everest. It can be described but not yet explained.

This book has been one long, sustained effort to shed redundant concepts. We now are down to two: a static but well-behaved wave function and the configuration space. The latter is Platonia, our pitch. I look at it as a child might – what a lopsided thing it is! However I turn in my mind the notion of 'thing', the space of all things constructed according to one rule comes out asymmetric. All the mathematical structures built by physicists to model the world have this inherent asymmetry. One rule creates triangles, but they are all different. No matter how you arrange them, their configuration space falls out oddly. Have another look at Triangle Land (Figures 3 and 4), which is just about the simplest Platonia there can be. And what does it look like? An upturned Matterhorn. Imagine trying to play football on that pitch.

The barest arena in which we can hope to represent appearances is the set of possible things. If we banish things, we banish all the world. So we have kept things and made them the instants of time. As we experience them, they are invariably time capsules. This is the principal contingent fact of existence: the wave function of the universe, playing the great game in timelessness, seeks and finds time capsules. What all-pervasive influence can put such a rooted bias into the game? The explanation seems to scream at us. Platonia is a skewed continent.

My conjecture is this. The Wheeler–DeWitt equation of our universe concentrates any of its well-behaved solutions on time capsules. I suspect the same result would hold for many different equations and configuration spaces. The inherent asymmetry of the configuration space will always 'funnel' the wave function onto time capsules.

I could fill up pages with hand-waving arguments for why this should be so, but they would baffle the non-specialist and offend the specialist. I shall attempt only to show that the 'seeking out' of time capsules need not depend on time and a special initial condition. Stationary equations may also do the trick. I may also mention that if, as explained in the

Notes, the universe can, in accordance with my recent insights, be understood solely in terms of pure structure, so that absolute distance plays no role, it will certainly be possible to make the arguments of the remaining parts of this chapter more precise and convincing.

TIMELESS DESCRIPTIONS OF DYNAMICS

In Part 2, we saw that Newtonian classical histories of the universe can be described in a timeless fashion as 'shortest' curves (geodesics) in configuration space. All histories that have the same energy can be described this way. This fact is often helpful, simplifying the solution of problems. In quantum mechanics something similar happens.

Physicists often want to know what will happen if some atomic particle is shot at a target of other particles. One way is to represent it by a 'cloud' of wave function – a wave packet – that moves towards the target. As Schrödinger showed, such a packet can be formed from waves corresponding to a relatively small range of momenta and energies, and it moves with a more or less definite velocity. When it reaches the target, it is 'scattered' – the wave function flies off in many directions. Physicists use the time-dependent Schrödinger equation to find the probabilities for these various directions. In this picture, the wave packet moves and is in different positions at different times. This is rather like the representation of history in Newtonian physics as an illuminated spot moving along a curve in configuration space.

There is, however, an alternative method. A single static wave covers the whole region – before, at, and after the target – traversed by the wave packet in the first picture. This one wave satisfies the stationary Schrödinger equation and corresponds to a particle with the average momentum and energy of the packet. As far as the target, the static wave is regular and plane, but at the target its pattern gets broken up. The interesting thing is that if we examine the pattern of the disrupted (but still static) wave in the region behind the target, we can deduce from it the probabilities with which the particle will be scattered in different directions in the first picture. I shall not go into details; suffice it to say that the one static wave is a kind of record of all the successive wave-packet positions in the first description. This is closely analogous to the way in classical physics in which the curve in the configuration space is a summary of all positions of the illuminated spot taken to represent the system at different times.

Interestingly, Max Born made his pioneering scattering calculations in the newly created wave mechanics by the second method. At that time Schrödinger had not even published his time-dependent equation. All the great early discoveries in wave mechanics, including Born's statistical interpretation of the wave function (which he came to by mulling over his scattering calculations), were made before the supposedly more fundamental and 'correct' time-dependent equation had been found. I find this suggestive. It strengthens my belief that all the physics of the universe can be described by a timeless wave equation. In fact, Mott also used the stationary equation to obtain the alpha-particle tracks. That timeless equation can locate time capsules.

But 'can' is not 'must'. The fact is that Mott used a special technique, always followed in such calculations, that mimics the wave-packet behaviour. The answer is to some extent simply assumed rather than truly derived and demonstrated. This can be done because the time-dependent and stationary Schrödinger equations have different structures, the latter having an extra freedom not present in the former. At each stage of his calculations, Mott systematically exploited this extra degree by making a definite kind of choice. This choice was not imposed by the mathematics but was made, probably instinctively, to match his temporal intuition. In fact, Mott's solution is not a proper solution at all but a kind of book-keeping record of how the real process would unfold in time. In addition, the condition corresponding to low entropy was also assumed rather than derived.

My conjecture seems to rest on a shaky basis. But there is more than one way of looking at this. The arguments for a timeless quantum universe are strong. The timelessness of the Wheeler–DeWitt equation, found by well-tried quantization methods, reflects the deepest structure of Einstein's theory. Quite independently, we never observe anything other than time capsules – the entire observable universe is marked, at all epochs, by profound temporal asymmetry. If we trust the equation, our observations tell us the outcome of a mathematical calculation performed by the universe itself using that equation. For that is what the contingent universe must be: a solution of the equation. If what we observe – a profusion of time capsules – is a representative fact, then the equation does concentrate Ψ on time capsules.

We can take Mott's solution more seriously. Several points can be made. Situations in which part of a quantum system is in the semiclassical regime, so that Hamilton's 'light rays' are present as latent or even incipient classical histories, are rather common and characteristic. The Heisenberg–Mott work then shows that such latent histories will become entangled with the remaining quantum variables, which must, in some way, reflect and carry information about those histories. What is not clear is whether the histories will exhibit a pronounced sense of direction – an arrow of time. That, above all, is put into the Mott solution by hand.

Also relevant is the mathematically somewhat suspect procedure known as successive approximation used to construct the Mott solution. There is no global arena in which the cloud chamber resides. Its atoms are effectively located in empty Euclidean space, and Mott could keep on adding approximations without worrying about their behaviour far from the cloud chamber. He was not constructing a genuine well-behaved solution, in which one must ensure the behaviour is right everywhere, especially at infinity. Instead, Mott used infinity as a kind of dustbin. This could not be done in a realistic situation, as I would now like to show.

A QUANTUM ORIGIN OF THE UNIVERSE?

When Planck made the first quantum discovery, he noted an interesting fact. The speed of light, Newton's gravitational constant, and Planck's constant clearly reflect fundamental properties of the world. From them it is possible to derive the characteristic mass m_{Planck}, length l_{Planck} and time t_{Planck}, with approximate values

$$l_{Planck} = 10^{-33} \text{ centimetres}$$
$$t_{Planck} = 10^{-43} \text{ seconds}$$
$$m_{Planck} = 10^{-5} \text{ grams}$$

On atomic scales the Planck mass is huge, corresponding to about 10^{19} hydrogen atoms. In contrast, the Planck length and time are far smaller than anything physicists can currently measure.

Much of current cosmology is concerned with the 'interface' of quantum gravity and classical physics. The universe around us is described by general relativity. This classical treatment is said to be valid right back into the distant past, very close to the Big Bang. The quantum phase of cosmology is supposed to become important only at extraordinarily small scales, of the order of the Planck length, 10^{-33} centimetres. Lights travel this distance in 10^{-43} seconds, and it is argued that quantum gravity 'comes into its own' only in this almost incomprehensibly early epoch.

All researchers agree that the nature of reality changes qualitatively in this domain. Different laws must be used. Time ceases to be an appropriate concept: things do not become, they are. In a process often likened to radioactive decay, our classical universe that emerges at the Big Bang is represented as somehow 'springing' out of timelessness, or even nothing. A mysterious quantum birth creates the initial conditions that apply at the start of the classical evolution. Our present universe is then the outcome of the conditions created by quantum gravity. The dichotomy between the laws of nature and initial conditions is thus resolved if the quantum creation process can be uniquely determined.

Stephen Hawking has long been working on this problem, and believes it can be solved by his so-called no-boundary proposal, a mechanism which should lead to a unique prediction for the initial conditions. His 'imaginary-time' mechanism, described in his *A Brief History of Time*, seemed to have the potential to do this. However, it has been widely criticized, and there are technical problems. The most serious seems to be that even if the mechanism can be made to work it will not produce unique initial conditions. Where Hawking has led, many have followed, and numerous creation schemes have been proposed.

My difficulty with this approach is the division introduced between the quantum and classical domains. One could almost get the impression that the laws of nature actually change, and I am sure that no theoretical physicist believes that. The approach is adopted because the physical conditions are hugely different in the two domains. In physics it is very common to use quite different schemes if the conditions studied are different. No engineer would use quantum mechanics to describe water flow in pipes, for example. But the much more appropriate hydrodynamic equations are consequences of the deeper quantum equations, and are valid in the appropriate domain.

Cosmology may be different. Most physicists have a deeply rooted notion of causality: explanations for the present must be sought in the past (vertical causality, as I have called it). This instinctive approach will

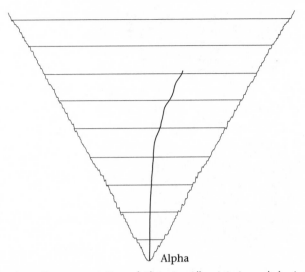

Alpha

Figure 53 A schematic representation of Platonia. All points in each horizontal section represent configurations of the universe with the same volume but different curvatures and matter distributions in them. According to the ideas of quantum creation, as yet unknown laws of quantum gravity hold near Alpha, and in some rather mysterious way give rise to conditions under which our universe – and with it, time – 'spring out of Alpha'. The thread shown ascending from Alpha represents the history of our universe that results from the enigmatic quantum creation.

be flawed if the very concept of the past is suspect. If quantum cosmology really is timeless, our notion of causality may have to be changed radically. We cannot look to a past to explain what we find around us. The here and now arises not from a past, but from the totality of things (horizontal causality).

Figure 53, a schematic representation of Platonia, may help. This is the skewed continent, as the cone shape makes clear. The quantum-creation approaches imply that the enigmatic and as yet unknown laws of quantum gravity create at the vertex – Alpha – a 'spark in eternity'. The spark, in its turn, creates close to Alpha the initial conditions of our actual universe. Time is born at the 'spark'. Our classical universe is the thread ascending through Platonia to our present location.

Figure 54 shows what is often called the *chronology of the universe* (the vertical axis is time, and the 'quantum creation' at Alpha occurs at the bottom left). Here each horizontal section is one point on the thread through Platonia in Figure 53: it is space at the corresponding time. The characteristic structures in space at the various cosmic epochs are shown:

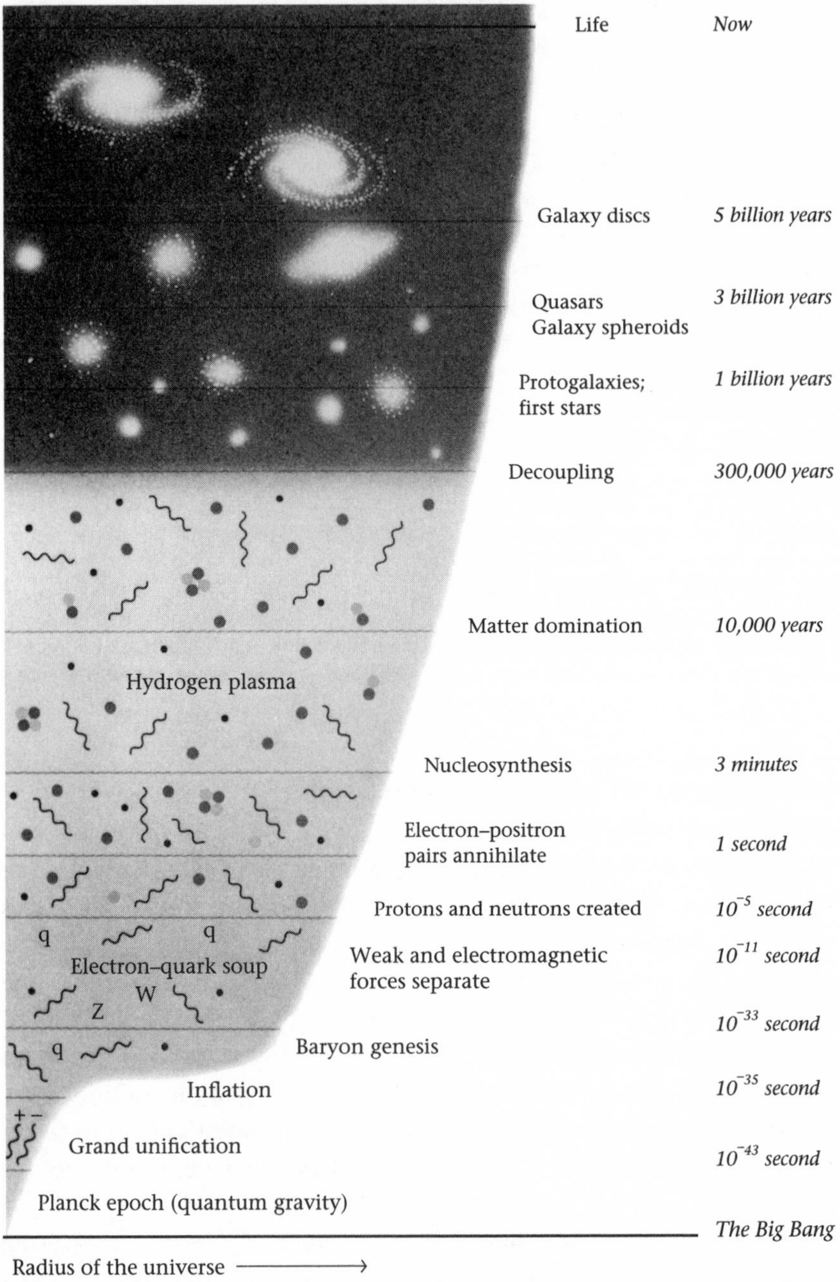

Life	Now
Galaxy discs	5 billion years
Quasars Galaxy spheroids	3 billion years
Protogalaxies; first stars	1 billion years
Decoupling	300,000 years
Matter domination	10,000 years
Nucleosynthesis	3 minutes
Electron–positron pairs annihilate	1 second
Protons and neutrons created	10^{-5} second
Weak and electromagnetic forces separate	10^{-11} second
	10^{-33} second
Baryon genesis	
Inflation	10^{-35} second
Grand unification	10^{-43} second
Planck epoch (quantum gravity)	
	The Big Bang

Hydrogen plasma

Electron–quark soup

Radius of the universe ⟶

Figure 54 Chronology of the universe. Redrawn from *A Short History of the Universe* by Joseph Silk (W. H. Freeman/Scientific American Library, 1994). The vertical axis represents time. Alpha, the 'quantum creation', occurs at the bottom left.

quarks in a soup near Alpha, primordial hydrogen and helium after the first three minutes, incipient galaxies a few thousand years later, and so on, right through to life on Earth at the present. Such is the thread 'born in the quantum spark'. The Everettian quantum cosmologists believe that the one quantum spark creates many such threads, one of them ours. But is there a 'spark' at Alpha? The laws of quantum gravity hold not just near Alpha, but throughout Platonia – they are the conjectured universal and ultimate laws. We have to ask what kind of solutions they can have, and how the solutions are created. How do threads, one or many, emerge?

Our entire experience tells us that the well-behaved solutions of the stationary Schrödinger equation that describe the characteristic structures of atoms, molecules and solids are determined by the complete structure of the configuration spaces on which they are defined. They exhibit global sensitivity – their behaviour has to be right everywhere. Since the governing equation does not contain the time, this delicate 'testing out of all possible behaviours' takes place in timelessness. If the Wheeler–DeWitt equation is like the stationary Schrödinger equation, then Alpha, where time is allegedly born, plays an important role, but it is not the locus at which some all-decisive die is cast. Of course it is singular – Platonia abuts on nothing at Alpha – but there are innumerable other special points scattered all over Platonia. None are quite like Alpha but, together with the overall shape of Platonia, they all have their role to play. Quantum mechanics is nothing if not democratic. Solutions of the Wheeler–DeWitt equation must be produced by a kind of dialogue between every point in Platonia.

The picture suggested by such arguments is this. Platonia as a whole determines how the static wave function 'beds down' on its landscape. There will be regions of semiclassical behaviour with respect to a large set of macroscopic quantities in which latent classical histories are defined. Just as the latent alpha-particle tracks get correlated through the wave function with the chamber electrons, 'nudging' the wave function onto time capsules, the same thing can happen in Platonia.

VISION OF A TIMELESS UNIVERSE

Let me now give you my vision of quantum cosmology, contrasting it with the chronology of the universe (Figure 54), that temporal representation of the 'thread' in Figure 53. Every instant of time you can conceive of is somewhere in Platonia. But the instants of time can themselves be richly structured beyond imagination. All things we see around us now in the universe are just parts of instants of time. All over Platonia there exist instants of time in which Wagner is composing *Tristan and Isolde*, astronauts are repairing the Hubble Space Telescope, birds are building nests and I am baking bread. The wave function of the universe finds its way to very few of them. The structure of the wave function and the form of the laws of nature – in which the tendency of gravity to clump matter is surely vital – forces the blue mist to seek out the most special instants, strung out along delicate threads. I think it is wrong of cosmologists to call Figure 54 a chronology of the universe. It is the map of a footpath in Platonia. The blue mist shines at instants containing time capsules, all of which, in their different ways, tell stories of a journey from Alpha along a fine thread of 'history' – a path winding through Platonia. Time is in such instants since they reflect the story of the path and, since the structure of Platonia in its totality forces the universal wave function to 'light up' the paths, there is a sense in which these instants reflect everything that is.

However, whereas alpha particles create, through their tracks, a literal image of history, the time capsules of the real universe embody their stories in a much subtler manner. This is inevitable given the grandeur of the story – cosmology in its entirety. Consider, for example, the Sun. Quantum mechanically, it will need to be represented in a configuration space of, say, 10^{60} dimensions, but vast stretches of it will be virtually devoid of wave function. The mere fact that the Sun is roughly spherical and can be well modelled by the laws of stellar structure sweeps most of the configuration space clean of wave function. The particular abundances of the chemical elements within the Sun have the same effect, drastically limiting the region of the solar configuration space in which the blue mist is concentrated.

One configuration at which the blue mist does shine brightly will be a characteristic distribution of all the particles in the Sun. To an experienced astrophysicist, this distribution tells an immensely rich story stretching back to the first three minutes (in the standard picture) when the primordial hydrogen and helium abundances were established. The whole story of the cosmos that we call our own is written in the distribution of the Sun's particles: the formation of galaxies and the earliest generations of stars; the supernova explosion that triggered the formation of the Sun and the solar system, and left the radioactivity that still powers so much tectonic and volcanic activity on the Earth; and the Sun's steady burning of its nuclear fuel.

The decisive element in this picture is the seed – or rather, seeds – from which these stories can all grow by the penetration of the wave function into the nooks and crannies in Platonia where the configurations are coherent stories. The wave function can be present there only if it is entangled with the latent histories of a semiclassical wave function established at least somewhere in Platonia. These are the Hamiltonian 'light rays' from which everything must 'grow'. Where are they likely to run, and what will their properties be? This is where the shape of Platonia must become decisive. The points in Platonia near Alpha containing Wagner and kingfishers are simply not visited by the blue mist, since Platonia as a whole lays out the latent histories in patterns that do not get entangled with such points.

Modern classical cosmology gives some hints as to where the latent histories might run. The simplest Big Bang cosmological solutions of Einstein's equations, first discovered by the Russian mathematician Alexander Friedmann in the early 1920s, have the maximum degree of symmetry and therefore ascend the central line in Figure 53. This is the history distinguished in cosmology by the universe's contents, despite the relativity of simultaneity. The universe explodes out of the singular state of zero volume, expands to a maximum volume and then recontracts to zero volume, gravity having halted and then reversed the initial, very rapid expansion. As shown schematically in Figure 55 (see p. 321), the universe ends in a Big Crunch. In other cosmological models, which normally require the universe to be spatially infinite, the expansion is so violent that the expansion is never halted.

In reality the universe is not completely symmetric, and the path out from Alpha is not exactly retraced, as shown in Figure 55, in which the rays emanating from Alpha are a measure of the relative 'irregularity' of the spatial configurations of the universe. Up the vertical ray, the universe is perfectly smooth, but on the rays that fan out at progressively larger angles the relative irregularity increases. The diagram shows one classical history which, at one end, starts in an almost perfectly smooth state but then becomes more and more irregular, due to the formation of galaxies, stars, black holes, planets and even human beings. This history reaches maximum expansion, turns round and recontracts, becoming more irregular all the time. It returns to the state of very small volume at a different point of Platonia, since although the volumes are small in both cases there are many additional variables that describe the structure of the state. Thus Alpha, the 'end of Platonia', is not a true point but actually a huge space of different possibilities, all with vanishingly small volumes.

Regarded purely as a path through Platonia, we cannot say that one end of this history is its beginning and the other is its end. I have lapsed into conventional talk. Such a priori notions do not belong in a timeless theory. Nevertheless, the two ends of the path are very different in nature, and it is tempting to say that, if our own existence is associated with such a path, its smooth end is what we would call the past and the irregular end the future. This would be very much in the tradition, initiated by Boltzmann, of suggesting that our sense of the forward flow of time, its arrow, is grounded solely in the increase in disorder that virtually all classical trajectories must exhibit if they pass through an exceptionally ordered region. Normally there are trajectories that both enter and leave such regions, and the entropic explanation of the arrow of time suggests that time will seem to flow forward in both directions out of these regions. In the present example, the exceptional region is on the frontier of Platonia, so the path truly ends there. This is beginning to look quite promising as the basis for a total explanation of time, but several conditions must be met. Before we address them, it is worth saying a little about gravity and thermodynamics.

The central conclusion of standard thermodynamics with an external time is that, if the low entropy of the world and its habitual increase are to be explained, the universe must presently be evolving out of a statistically most unlikely state. In systems in which gravity does not act, the unlikely state is generally one that is structured, while the likely state is characterized by a bland uniformity. Gas confined in a finite volume tends quickly to a very uniform state in which it occupies all the available space and all temperature differences are levelled out. This is the equilibrium state. It is vastly more probable than any ordered state because there are so many more ways in which it can be realized microscopically. The situation is much more complicated when gravity comes into play, since there is no well-defined equilibrium state for a gravitating system. Gravity is attractive, so a uniform state is unstable and will tend to break into self-gravitating clumps. This is the exact opposite of a gas.

Currently there is no fully satisfactory thermodynamics of cosmology, mainly because of the way in which gravity acts. But it does seem certain that black holes, which almost certainly exist, have a well-defined entropy associated with them. This was the final and most dramatic outcome of the intensely exciting 'golden decade' in the study of black holes that ended with the discovery of black-hole evaporation by Stephen Hawking in 1974. This fascinating story has been told with great verve by Kip Thorne in his *Black Holes and Time Warps*. The entropy associated with black holes is staggeringly large. Since there is little evidence that black holes existed at very early times but a lot that many have since been formed and more will be formed, the universe seems to have begun in an extraordinarily unlikely state.

No one has done more than Roger Penrose to highlight this fact. His *The Emperor's New Mind* has an entertaining illustration of the creating divinity seeking with a pin to find the tiny improbable point of the initial condition of the universe from which its utterly unlikely history must have sprung. Penrose seeks to explain this in a theory in which both time and actual quantum-mechanical collapse are real, and the laws of nature are inherently asymmetric in time. My approach is quite different because I think that the whole problem of time and its arrow can – paradoxically – be formulated more precisely and transparently in a context in which time does not exist at all. I also believe that, far from being highly unlikely, the kind of history and cosmos we experience are characteristic and likely in a timeless scenario.

It all depends on how a static wave function 'beds down' on the starkly asymmetric continent of Platonia. The issue of the correct arena is all-important. The collective intuition of most physicists, wedded to time and honed on translucent structures like absolute space, is forced to see the observed universe as highly improbable. But in Platonia it may appear inevitable. Wave functions have a way of finding special structures: for example, they can create complex molecules like proteins and DNA.

Let it be granted, not unreasonably I think, that the wave function of the universe will be semiclassical with respect to at least some variables in some part of Platonia. Where is it likely to be, and how will the corresponding Hamiltonian 'light rays' run? Do they emerge from Alpha? Here, one of the most famous results of classical general relativity may be relevant. Penrose and Hawking showed that its solutions have a remarkable propensity to evolve into a singular state. All that is necessary is for sufficient matter to be concentrated within a certain finite region.

After that, as Penrose showed, collapse to a black hole is inevitable. Hawking showed that there is a sense in which the Big Bang itself can be regarded as the time-reverse of the Penrose collapse to a black hole. (Collapse here has nothing to do with quantum-mechanical collapse of the wave function.) Solutions that terminate at one or both ends in singular states are characteristic of general relativity.

What happens in a quantum theory cannot be totally unrelated to the corresponding classical theory. It therefore seems likely that in quantum gravity there will be a semiclassical region near the central ray in Figure 55. The Hamiltonian 'light rays' in it may well, reflecting the structure of Platonia, appear to emanate from Alpha, and rise up in a kind of jet which then spreads out and falls back, as in a fountain, returning to small volumes but in a much more irregular state. Alternatively, they may go on for ever, receding ever farther from Alpha. I have described these trajectories as if they were traced in time, but they are only paths.

Moreover, the paths are still only 'seeds'. The finding of the full rich structures which tell us so insistently that time exists and flows must result from entanglement with the host of the remaining quantum variables that constitute the expanses of Platonia. When discussing alpha-particle tracks, I emphasized that Mott employed a special device to concentrate the wave function on time capsules. Considered purely in terms of the stationary Schrödinger equation, this was artificial. This is what created the static alpha-particle tracks and such a strong sense of time and history out of the 'seed' of a spherical wave pattern.

If my proposal is along the right lines, there must be some natural and plausible mechanism within static quantum cosmology which performs this task. Platonia at large must force it to happen. As I have already said, I see the cause in the rooted asymmetry of contingent things. Platonia is necessarily skew. It is easy to imagine that the cone of Figure 55 'funnels entanglement outwards', much as a trumpeter blows air from a bugle. I deliberately chose this last simile. The bugle does create a nice image of what I have in mind, but it also creates hot air. There are no hard mathematical proofs to support my idea, but I hope you are now persuaded that at least the arguments for a timeless universe are strong. If it nevertheless appears intensely temporal, there must somewhere be a massive reason for the fact. I think it is the asymmetry of being. Being can be more or less. Sitting in the midst of things, we feel ourselves carried forward on the mighty arrow of time. But it is an arrow that does not move. It is

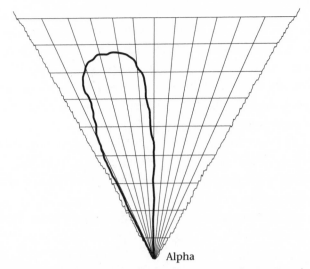

Alpha

Figure 55 Explosion out of the Big Bang and recontraction to the Big Crunch according to present standard cosmology.

simply an arrow that points from the simple to the complex, from less to more, most fundamentally of all from nothing to something. If we could look over our shoulder in Platonia, we should see where this trek began: at the edge of nothing.

A WELL-ORDERED COSMOS?

Let me end the main part of the book with a few comments on structure, and what strikes a theoretical physicist as improbable. If we think that dynamical histories in space and time are the fundamental things in nature, then all statistical reflections on the world lead to great difficulties. Most histories are unutterably boring over all but a minuscule fraction of their length. We can never understand the miracle of the structured world. Things are completely changed if quantum cosmology is really about some well-behaved distribution of a static wave function over Platonia. The configurations at which Ψ collects strongly must be special – in some sense they must resonate with all the other configurations that are competing for wave function. Quantum cosmology becomes a kind of beauty contest in eternity. The winners – those that get a high probability density – must be exceptional, like the DNA molecule. This is just the opposite of what classical physics leads us to expect. There, the winners are boring.

When I wrote my book on the history of dynamics, I was exposed to the beautiful poetic notion of the well-ordered cosmos. Intermittent reading of Leibniz had already made me deeply interested in structure, and this was greatly strengthened by my collaboration with Lee Smolin on the Leibnizian idea that the actual universe is more varied than any other conceivable universe. That still remains a mere idea (though Paul Davies was sufficiently intrigued to include a brief account of the idea in his *The Mind of God*), but I became persuaded that a scientific theory of the universe in which structure is created as a first principle is possible. We may need to get back to the wonder of childhood to comprehend what the world really is. Yeats once wrote of Bishop Berkeley that he 'has brought back to us the world that only exists because it shines and sounds. A child, smothering its laughter because the elders are standing around, has opened once more the great box of toys.'

The single most striking thing about the universe we see around us is its rich structure, which is so difficult to understand on a priori statistical grounds. Until the modern scientific age, all thinkers saw the first task of science as being the direct description and explanation of this structure. This natural impulse is reflected in the Pythagorean notion of the well-ordered cosmos. It was still very strong in both Kepler and Galileo. However, when Newton demonstrated the supreme importance of accelerations in dynamics, the perspective of science changed, for the world at the present instant became the mere consequence of its initial conditions. Instead of asking directly how structure is fashioned, science turned to asking how it is *re*fashioned.

It seems to me that the abolition of time in quantum gravity must bring us back to a more Pythagorean perspective, though with a quantum slant, for now we must simply ask what structures are probable. It seems to me that the first decisive step in this direction was the discovery by Schrödinger of his time-independent wave equation, with its all-important condition that its solutions must be well behaved, and Born's probability interpretation of quantum mechanics. For we know that these two basic elements of quantum mechanics work together to bring forth exquisite structures in great profusion, doing so moreover without any boundary or initial conditions and with total disregard for what might seem statistically likely. That is the story of atomic, molecular and solid-state physics. I think it may even be the story of the universe.

Life Without Time

Pied Beauty

Glory be to God for dappled things –
For skies of couple-colour as a brinded cow;
For rose-moles all in stipple upon trout that swim;
Fresh-firecoal chestnut-falls; finches' wings;
Landscape plotted and pieced – fold, fallow, and plough;
And áll trádes, their gear and tackle and trim.

All things counter, original, spare, strange;
Whatever is fickle, freckled (who knows how?)
With swift, slow; sweet, sour; adazzle, dim;
He fathers-forth whose beauty is past change:
Praise him

It is a pity that Gerard Manley Hopkins's finest line in this poem implies that creation is a male prerogative and is so inappropriate for the dawning millennium. But what beauty past change the wave function does manage to find in the nooks and crannies of Platonia! What are we to think of life if time and motion are nothing but very well-founded illusions? I have selected a few topics, trying to anticipate some of the questions that the reader, as a human being rather than a scientist, might ask. I also give some hints of how I think the divide between impersonal science and the world of the arts, emotions and religious aspirations might be bridged somewhere in Platonia. I love and respect the disciplines of both. Can they be shown to flow from a common view of the world?

Can We Really Believe in Many Worlds?

The evidence for them is strong. The history of science shows that physicists have tended to be wrong when they have not believed counter-intuitive results of good theories. However, despite strong intellectual acceptance of many worlds, I live my life as if it were unique. You might call me a somewhat apologetic 'many-worlder'! There are occasions when the real existence of other worlds, other outcomes, seems very hard to accept. Soon after I started writing this book, Princess Diana was killed, and Britain – like much of the world – was gripped by a most extraordinary mood. Watching the funeral service live, I did wonder how seriously one can take a theory which suggests that she survived the crash in other worlds. Death appears so final.

Such doubts may arise from the extraordinary creative power – whatever it is – that lies behind the world. What we experience in any instant always appears to be embedded in a rich and coherent story. That is what makes it seem unique. I would be reassured if the blue mist did indeed seek out only such stories. Shakespeare wrote many plays, nearly all masterpieces. But we do not even have a unique *Hamlet*: producers are always cutting different lines, and producing the play in novel ways. Variety is no bad thing: I have enjoyed many outstanding *Hamlet*s. In the timeless many-instants interpretation, they were all other worlds, and that is what makes timeless quantum cosmology fascinating. Our past is just another world. This is the message that quantum mechanics and the deep timeless structure of general relativity seem to be telling us. If you accept that you experienced this morning, that commits you to other worlds. All the instants we have experienced are other worlds, for they are not the one we are in now. Can we then deny the existence of worlds on which Ψ collects just as strongly as on our remembered experiences?

Does Free Will Exist?

Anyone committed to science has difficulty with free will. In *The Selfish Gene* (2nd edition, pp.270–71), Dawkins asks, 'What on earth do you think you are, if not a robot, albeit a very complicated one?' From personal introspection, I do not believe that my conscious self exercises free will. Certainly I ponder difficult decisions at length, but the decision itself invariably comes into consciousness from a different, unconscious realm. Brain research confirms that what we think are spontaneous decisions, acts of free will, are prepared in the unconscious mind before we become aware of them.

However, the many-instants interpretation puts an intriguingly different slant on causality, suggesting that it operates in nothing like the way we normally believe it to. In both classical physics and Everett's original scheme, what happens now is the consequence of the past. But with many instants, each Now 'competes' with all other Nows in a timeless beauty contest to win the highest probability. The ability of each Now to 'resonate' with the other Nows is what counts. Its chance to exist is determined by what it is in itself. The structure of things is the determining power in a timeless world.

The same applies to us, for our conscious instants are embedded in the Nows. The probability of us experiencing ourselves doing something is just the sum of the probabilities for all the different Nows in which that experience is embedded. Everything we experience is brought into existence by being what it is. Our very nature determines whether we shall or shall not be. I find that consoling. We are because of what we are. Our existence is determined by the way we relate to (or resonate with) everything else that can be. Although Darwinism is a marvellous theory, and I greatly admire and respect Richard Dawkins's writings, one day the theory of evolution will be subsumed in a greater scheme, just as Newtonian mechanics was subsumed in relativity without in any way ceasing to be great and valid science. For this reason, and for the remarks just made, I do not think that we are robots or that anything happens by chance. That view arises because we do not have a large enough perspective on things. We are the answers to the question of what can be maximally sensitive to the totality of what is possible. That is quite Darwinian. Species, ultimately genes, exist only if they fit in an environment. Platonia is the ultimate environment.

In Box 3, I said that Platonia is a 'heavenly vault' in which the music of the spheres is played. This formulation grew out of numerous discussions with the Celtic composer, musicologist and poet John Purser (brother of the mathematician and cryptographer Michael, who made the comment about my parents with which I ended the Preface, and brought to my attention the Shakespeare quotation). With the inimitable assurance of which only he is capable, John is adamant that the only theory of the universe that ever made sense was (is) the music of the spheres. My guts tell me that he and the artists quite generally are right. But harmony rests on mathematics, of course. Rather appropriately, given my extensive use of meteorological metaphor, John and his wife Bar live in the misty Isle of

Skye, where at least one of the said discussions took place while the better part of a bottle of whisky was consumed, mostly by John.

You will naturally ask why we do not hear this music of the spheres. Keats provides a first answer: 'Heard melodies are sweet, but those unheard Are sweeter'. But Leibniz may have given the true answer. In his monadology, he teaches that the quintessential you, everything you experience in consciousness and the unconscious, is precisely this music. You are the music of the spheres heard from the particular vantage point that is you. This is taking a little liberty with the letter but certainly not the spirit of his great philosophical scheme. On the subject of liberties, I have taken fewer with Leibniz than Michael did with Shakespeare. Hal does not actually ask Falstaff ('fat-witted with drinking of old sack') why he should be so superfluous 'to inquire the nature of time' but to 'demand the time of day'. But, were it not for the blessed Sun and its diurnal rotation (our fortunate circumstances), the one question would be as profound as the other.

Now for something like the original *Gretchenfrage*.

Is There a Role for a Creator in Quantum Cosmology?

Perhaps, but it is a somewhat strange one. It seems to me that science can never do more than guess – theorize about – the structure of things and then test to see whether its conjectures are confirmed. This is an open-ended venture (with tremendous successes behind it) and always presupposes that there is some structure already out there waiting to be found. In the scheme I have advanced, much is presupposed: Platonia, its detailed structure (immensely important) and a wave function that 'samples' possibilities. It is the nature of theory to presuppose something, so that always leaves a potential role for a Creator. But does invoking something to explain what we cannot explain get us any further?

What does intrigue me is the power of structures in a timeless scheme. They determine where the wave function collects. If one wanted to see Ψ as spirit pondering what shall be brought into existence, it has no power in the matter. Leibniz always said that not even God could escape the dictates of reason. He must always act rationally. Perhaps that is more reassuring than a capricious deity is. However, a rational universe is quite alarming too. If you are about to perish in a concentration camp, is it any consolation to know that what must be will be?

In *The Life of the Cosmos*, my friend Lee Smolin espouses a self-creating universe, likening its growth to the often largely unplanned development of cities. I find his epilogue especially eloquent. In fact, timeless quantum cosmology does give almost god-like power to structures, ourselves included, to bring themselves into being. We shall be if that fits the great scheme of things. The ideas of both Lee and myself tend to pantheism. The whole universe – Platonia *and* the wave function – is the closest we can get to a God.

Where Is Heaven?

I have long thought that, if only we had the wit to see it, we are already in heaven. It is Platonia. I say this with some trepidation, though I believe it is true. If so, Platonia must be hell and purgatory as well. What I mean by this is really quite simple: some places in Platonia are very admirable, pleasant and beautiful, many are boring in the extreme, and others are horrendously nasty. The same contrasts exist within the individual Nows. What we do not know is where the wave function collects.

I certainly find it difficult to believe that there is a material world in which we currently find ourselves, and some other, quite different, immaterial world we enter after death. Apart from anything else, modern physics suggests very strongly that so-called gross matter – the clay from which we are made – is anything but that. It is almost positively immaterial. Platonic forms have exact mathematical properties, and those are all that physicists need to model the world and to attribute to matter.

I also feel strongly that *this* created world is something to be marvelled at and cherished, not dismissed as some second-best version of what is yet to come. Disrespect for this world is disrespect for whatever creates it. I shall not attempt to argue about these things in detail here, but the total elimination of time, if accepted and supported by mathematics and observation, must force theologians to reconsider their notions. If there is a happier and more perfect world, in which the lion lies down with the lamb and the sword is made into a ploughshare, I think it will simply be somewhere else in Platonia. I am sure that there are locations where experience is much deeper and richer than here. Such experience may be perfectly timeless – consciousness just sees what is. Perhaps we are somehow included in that awareness. Perhaps too the world is redeemed, and its inner conflicts resolved and understood somewhere in Platonia's distant reaches, farther from Alpha than we are.

It is not for nothing that I emphasized in the early part of the book that Platonia has an Alpha but no Omega. The idea of a Point Omega was introduced by the Jesuit biologist Teilhard de Chardin, who conceived of it as some kind of consummation of evolution in the ultimate future, 'on the *boundary* of all future time'. I have quoted these last words from Frank Tipler's book *The Physics of Immortality*, in which he argues that Point Omega is where our material universe recontracts to the Big Crunch. By then, he argues, intelligences will have become so adept and computerized that we shall all be recreated as virtual computer programs, run so fast that we have an effective eternity of existence in which we are resurrected before the universe ends in the Crunch.

I can only say that is not how I see things. I search in vain for Omega in Platonia and find only Alpha. But Platonia is a vast land. Let us cherish everything around us wherever we happen to find ourselves in the Platonic palace.

Of one thing I feel very sure. Many poets and theologians give a misleading image of heaven and eternity. Consider the opening lines of Vaughan's famous poem 'The World':

> I saw Eternity the other night
> Like a great Ring of pure and endless light,
>   All calm as it was bright;
> And round beneath it, Time, in hours, days, years,
>   Driven by the spheres,
> Like a vast shadow moved, in which the world
>   And all her train were hurled.

This is magnificent poetry as a description of the mystical state in which mind comprehends all structure as unified. But the real wonder of unity is when it knits together rich variety. Bliss is comprehension of many things in harmony at once. I do not think that eternity is pure and endless light. Such light merely illuminates, for example, the millions of leaves of the forest of the American fall when we see them all at once.

Is Time Travel Possible?

Time travel of a sort is possible within general relativity as a classical theory, but is subject to strict limitations. You cannot travel back into the past and kill your parents before your conception. In quantum cosmology, you can travel back to a parallel universe, and there kill your parents before they conceive you. However, we have to be careful about the use of 'you'. The person who 'travels' to these other worlds is not exactly you now. As the discussion of the haemoglobin molecule showed, the change within our bodies from one instant to another is stupendous. The fact that we have such an enduring sense of deep continuity of our personal identity is very remarkable. I see it as another manifestation of the creative power that brings everything into existence. Stephen Hawking long suspected that even if time travel is logically possible, it will have a very low probability in quantum cosmology. That is my feeling too. Platonia certainly contains Nows in which there are beings whose memories tell them they have travelled backwards in time. However, I think such Nows have a very low probability.

To tell the truth, I find the idea of time travel boring compared with the reality of our normal existence. Each time capsule that represents an experienced Now reflects innumerable other Nows all over Platonia, some of them vividly. In a very real sense, our memories make us present in what we call the past, and our anticipations give us a foretaste of what we call the future. Why do we need time machines if our very existence is a kind of being present everywhere in what can be? This is very Leibnizian. We are all part of one another, and we are each just the totality of things seen from our own viewpoint.

Doesn't the Denial of Motion Take All Joy and Verve out of Life?

I do feel this issue keenly. The kingfisher parable should make that clear. In principle, there is no reason why we should not attempt to put our very direct sense of change directly into the foundations of physics. There is a long tradition, going back at least to Hamilton, that seeks to make *process* the most basic thing in the world. Roughly, the idea is that physics should be built up using verbs, not nouns. In 1929 the English philosopher Alfred North Whitehead published an unreadable – in my experience – book called *Process and Reality* in which he advocated process. It all sounds very exciting, but I just do not think it can be done, despite a valiant attempt by Abner Shimony. Having translated seventy million words of Russian into English, I can say with some feeling that sentences do have a subject and generally an object. I could have written this book using the one verb 'to be', which hardly counts as a verb. For this reason it seldom appears in Russian; when it does, it is most often as a surrogate: 'to appear'. But a book without nouns is nothing. Not even James Joyce could write it. For some reason, disembodied verbs exert a fascination not unlike the grin of the Cheshire cat. But when Owen Glendower claimed to be able to 'call up spirits from the vasty deep', Hotspur answered: 'Aye, and so can I and any man, but will they come when you call them?' I should like to see it done.

Less provocatively, I wonder if, at root, there is that much difference between the Heraclitan and Parmenidean schools, representing 'verbs' and 'nouns' respectively. If my definition of an instant of time is accepted, it becomes hard to say in what respect those two great Pre-Socratics might differ. The two best-known sayings attributed to Heraclitus are 'Everything flows' (*Panta rei*) and the very sentence which, entirely unconsciously, I used to clinch the argument that the cat Lucy who leapt to catch the swift was not the cat who landed with her prey: 'One cannot step into the same river twice.' There is always change from one instant to another – no two are alike. But that is just what I have tried to capture with the notion of Platonia as the collection of all distinct instants. Heraclitus argued that the appearance of permanence, of enduring substance, is an illusion created by the laws that govern change. Obviously, he and Parmenides could not be expected to have anticipated quantum mechanics, wave functions and the Wheeler-DeWitt equation, But I see Bell's account of alpha-particle track formation as remarkable support for the Heraclitan standpoint that appearances are the outcome of the laws that bring them forth. Would it not be a wonderful reconciliation of opposites if the static wave function were to settle spontaneously on time capsules that are redolent of both flux (evidence of history) and stasis (evidence that things endured through it)?

But the loss of motion is still poignant, a premonition of mortality and a view of our life from outside it. There is a scene in George Eliot's *Middlemarch* in which Ladislaw and a painter friend chance to see the heroine Dorothea in a particularly striking pose in Rome. The painter is keen to capture it on canvas, but Ladislaw taunts him:

> Painting stares at you with an insistent imperfection. I feel that especially about representations of women. As if a woman were a mere coloured

superficies! You must wait for movement and tone. There is a difference in their very breathing: they change from moment to moment.

Keats too, for all the beauty of his Grecian urn, addresses it with the words

Thou, silent form! Dost tease us out of thought
As doth eternity. Cold Pastoral!

When Keats wrote these lines, he must have known that all too soon his home would be a grave. Is Platonia a graveyard? Of a kind it undoubtedly is, but it is a heavenly vault. For it is more like a miraculous store of paintings by artists representing the entire range of abilities. The best pictures are those that somehow reflect one another. These are the paintings we find there in profusion. There are very few of the mediocre, dull ones. Despite what Ladislaw says, the best paintings have a tremendous vibrancy. Turner does almost bind you to the mast of the *Ariel*. Indeed, Ladislaw's own words immediately before the passage quoted above are: 'After all, the true seeing is within.' Frozen it may be, but Platonia is the demesne where 'Beauty is truth, truth beauty' and the boughs cannot shed their leaves 'nor ever bid the Spring adieu'. With that perfect ode, Keats did achieve the immortality for which he so desperately longed.

In a fine essay entitled 'The timeless world of a play', Tennessee Williams praises great sculpture because it

often follows the lines of the human body, yet the repose of great sculpture suddenly transmutes those human lines to something that has an absoluteness, a purity, a beauty, which would not be possible in a living mobile form.

He argues that a play can achieve the same effect, and so help us to escape the ravages of time. 'Whether or not we admit it to ourselves, we are all haunted by a truly awful sense of impermanence.' Again, this is very beautiful writing, but has Williams failed to see the truth all around us – the Platonic eternity we inhabit in each instant? Is it blindness that drives him to seek eternity? Some people can pass a cathedral without noticing it.

The desire for an afterlife is very understandable, but we may be looking for immortality in the wrong place. I mentioned Schrödinger's curious failure to recognize in his own physics the philosophy of ancient India (especially the Upanishads) he so admired. There is a beautiful passage at the end of his epilogue to *What is Life?* that nevertheless strikes me as wishful thinking. He poses the question, 'What is this "I"?' Here is part of his answer:

> If you analyse it closely you will, I think, find that it is just a little bit more than a collection of single data (experiences and memories), namely the canvas *upon which* they are collected. And you will, on close introspection, find that what you really mean by 'I' is that ground-stuff upon which they are collected. You may come to a distant country, lose sight of all your friends, may all but forget them; you acquire new friends, you share life with them as intensely as you ever did with your old ones ... Yet there has been no intermediate break, no death.

Thus, he argues that our personal 'ground-stuff' is imperishable, holding us together through all the changes of life. He ends with this affirmation of faith: 'In no case is there a loss of personal existence to deplore. Nor will there ever be.' But earlier he had praised the great Upanishads for their recognition that ATHMAN = BRAHMAN (the personal self equals the omnipresent, all-comprehending eternal self). He seems to want to have his cake and eat it, to be dissolved in the all-comprehending eternal self yet still retain a personal identity. He is not finding a canvas, he is clutching for a straw.

Tennessee Williams faced up to things more squarely: 'About their lives people ought to remember that when they are finished, everything in them will be contained in a marvelous state of repose which is the same as that which they unconsciously admired in drama.' Exactly: this is the truth and beauty of Keat's Grecian urn. Williams goes on: 'Snatching the eternal out of the desperately fleeting is the great magic trick of human existence.' Yes, though it's not a magic trick but simply the opening of our eyes.

Some years ago, I heard Dame Janet Baker interviewed on radio. She was asked if she ever listened to her recordings and, if so, what were her favourites. She said she almost never listened to them. For her, every Now was so exciting and new, it was a great mistake to try to repeat one. In her singing she made no attempt at all to recreate earlier performances and do the high points the same way as the night before. Again and again she spoke with the deepest reverence of the Now and how it should be new and happen spontaneously. 'The Now is what is real', she said.

I thought it was the perfect artistic expression of how I see timeless quantum cosmology. By definition, every Now in Platonia is new, for all the Nows are different. But some are vastly more interesting and exciting than others. Miraculously, these are the Nows that the wave function of the universe seems to find with unerring skill.

What a gift too is the specious present. Appreciation of poetry and music would be impossible without it. I can live without motion if I can sense it as the line that runs through a story all bound up in one Now. Janet Baker is right. Watching motion, listening to Beethoven, looking at a painting by Turner – all are given to us in the Now, which we experience as the specious present. Einstein seems to have regretted that modern science – and his own relativity in particular – had taken the Now, the vibrant present, out of the world. On the contrary, I think the Now may well constitute the very essence of the physical world, the first quantum concept (as David Deutsch refers to time). The artists always knew it was there, and worshipped at its altar. It was Dirac's rediscovery of the Now at the heart of general relativity that started my quest.

I do also feel that novelty is a genuine element of quantum mechanics, especially in the many-worlds form, not present in classical mechanics. In the main text I spoke of lying down to sleep and knowing 'not what we should wake to'. I see no fundamental line of time and causal evolution along which we march as robots; each experienced Now is new and distinct. I think that the many-worlds hypothesis is the scientific counterpart of the thrill of artistic creation that Janet Baker feels so strongly. It is something essentially new for which there is no adequate explanation in any supposed past from which we have tumbled via a computer algorithm. There is no explanation of any one triangle in terms of any others, and the same is true of all Nows.

The Italian painter Claudio Olivieri, a friend of Bruno Bertotti, creates paintings in which he evokes a sense of timelessness. He expresses his aim through this poem:

> È con la pittura che le apparenze si mutano in apparizioni:
> Ciò che è mostrato non è la verosimiglianza ma la nascita.
> È così che ci viene restituto il nostro presente,
> L'assolutamente unico ma imprevedibile presente,
> Somma di tutti tempi, raduno degli attimi che ci fanno
> Viventi, atto sempre inaugurale dell'esistere.

A free translation is as follows:

The painting transfigures semblance in sudden apparition,
showing not likeness but birth.
That is how we are given back our present,
The absolutely unique but unforeseeable present,
sum of all times, gathering of the moments that make us alive,
the ever inaugural act of existence.

This does express the main ideas I have tried to get across in the final part of the book. Each experienced instant is a separate creation (birth), the ever inaugural act of existence, brought to life by the gathering of all times. The thrill that Janet Baker experiences in each Now is the *assolutamente unico ma imprevedibile presente*, that finding of ourselves in one of the instants that quantum mechanics makes resonate especially strongly with other instants.

As I began, so I end. Turner has taught us the way to look at the world, and even how to come to terms with many worlds. Once any painting of his had reached a certain stage of completion, all additions to it became simply variations on an existing masterpiece. All the stages through which his paintings then passed were perfect, and each was – is – a separate world. Nature is an even more consummate artist than Turner. For he too is part of Nature. Turner is also right in the way he places us humans in the great arena. In nearly all his pictures, human beings, though tiny on the cosmic scale, are integral parts of some huge picture, Keats's urn painted large. We are simultaneously spectators and participants, subtly changing and constantly working on an inherited landscape. We are there in one place but bound up into something much larger. Gretchen Kubasiak gave me, besides the Tennessee Williams essay, some Aborigine philosophy that, but for the idea that we are visitors, chimes with this thought:

> We are all visitors to this time, this place. We are just passing through. Our purpose here is to observe, to learn, to grow, to love... And then we return home.

No, this is home. Mach once commented that 'In wishing to preserve our personal memories beyond death, we are behaving like the astute Eskimo, who refused with thanks the gift of immortality without his seals and walruses.' I am not going without them, either. I cannot even if I wanted to: they are part of me. Like you, I am nothing and yet everything. I am nothing because there is no personal canvas on which I am painted. I am everything because I am the universe seen from the point, unforeseeable because it is unique, that is me now. *C'est moi.* I am bound to stay. We all watch—and participate in—the great spectacle. Immortality is here. Our task is to recognize it. Some Nows are thrilling and beautiful beyond description. Being in them is the supreme gift.

NOTES

PREFACE

(1) (p. 2) The article about Dirac appeared in the *Süddeutsche Zeitung* for Friday, 18 October 1963, and was based on an article by Dirac that appeared in *Scientific American* in May 1963.

(2) (p. 4) On hearing about my plans for this book, Michael Purser brought to my attention the following rebuke from Prince Hal to Falstaff:

> Unless hours were cups of sack, and minutes capons, and clocks the tongues of bawds, and dials the signs of leaping-houses, and the blessed sun himself a fair hot wench in flame-colour'd taffeta, I see no reason why thou shouldst be so superfluous to inquire the nature of time.
>
> Henry IV, Part I (I. ii)
(I comment on this in the Epilogue.)

CHAPTER 1: THE MAIN PUZZLES

The Next Revolution in Physics (p. 14) The possible non-existence of time has just begun to be discussed in authoritative books for the general public. Both Paul Davies, in his *About Time,* and Kip Thorne, in his *Black Holes* and *Time Warps,* devote a few pages to the topic. In apocalyptic vein, Thorne likens the fate of space-time near a black hole singularity to

> a piece of wood impregnated with water . . . the wood represents space, the water represents time, and the two (wood and water, space and time) are tightly inter-woven, unified. The singularity and the laws of quantum gravity that rule it are like a fire into which the water-impregnated wood is thrown. The fire boils the water out of the wood, leaving the wood alone and vulnerable; in the singularity the laws of quantum gravity destroy time . . . (p. 477)

However, Thorne's magnificent book is devoted to other topics, and nothing prepares the reader for this dramatic and singular end of time. Moreover, the evidence, as I read it, is that timelessness permeates the whole universe, not just the vicinity of singularities. Paul Davies, for his part, repeatedly expresses a deep mystification about time. His book is almost a compendium of conundrums, and he candidly consoles the reader with 'you may well be even more confused about time after reading this book than you were before. That's all right; I was more confused myself after writing it' (p. 10). In fact, I think Paul's subtitle, *Einstein's Unfinished Revolution*, is the key to a lot of the puzzles. As we shall see in Part 3, there are aspects of physical time which Einstein did not address.

Among the popular books that I know, the two that undoubtedly give most prominence to the problem of time in quantum gravity are Lee Smolin's *The Life of the Cosmos*, which contains some discussion of my own ideas, and David Deutsch's *The Fabric of Reality*. There is considerable overlap between my book and Deutsch's chapter 'Time: the first quantum concept'. One technical book, now going into a third edition, that from the start has taken timelessness very seriously is Dieter Zeh's *The Physical Basis of the Direction of Time*.

It may be that the reason why a book like this one, devoted exclusively to the idea that time does not exist, has not hitherto been published by a physicist has a sociological explanation. For professionals working in institutes and dependent on the opinions of peers for research funding, such a book might damage their reputation and put further research in jeopardy. After all, at first it does seem outrageous to suggest that time does not exist. It may not be accidental that I, as an independent not reliant on conventional funding, have been prepared to 'come out'.

In this connection, my experience at a big international conference in Spain in 1991 devoted to the arrow of time was very interesting. The following is quoted from my paper in the conference proceedings (available in paperback as Halliwell *et al.*, 1994):

> During the Workshop, I conducted a very informal straw-poll, putting the following question to each of the 42 participants:
>
> *Do you believe time is a truly basic concept that must appear in the foundations of any theory of the world, or is it an effective concept that can be derived from more primitive notions in the same way that a notion of temperature can be recovered in statistical mechanics?*
>
> The results were as follows: 20 said there was no time at a fundamental level, 12 declared themselves to be undecided or wished to abstain, and 10 believed time did exist at the most basic level. However, among the 12 in the undecided/abstain column, 5 were sympathetic to or inclined to the belief that time should not appear at the most basic level of theory.

Thus, a clear majority doubted the existence of time. When I took my straw-poll, I said that I intended to publish the names with their opinions, which was why two people abstained, to remain anonymous. As it happens the conference generated immense media interest in Spain, not least because of the presence of Stephen Hawking and Nobel Laureate Murray Gell-Mann, and the reporter from *El Pais* got hold of a copy of my results. One of the participants (neither of the above), finding his own opinion quoted in a big article the day after the conference, was none too pleased and greeted me when we met six months later at a conference in Cincinnati with 'You and your damned straw-poll!' I then realized why the editors had meanwhile asked me to withhold the names in my paper, which I happily did.

It was at the later conference that I learned a *bon mot* of Mark Twain that somehow seems appropriate here: 'If the end of the world is nigh, it is time to be in Cincinnati. Everything comes to Cincinnati twenty years late.'

The Ultimate Things (p. 15) I mentioned in the Preface the difficulty of writing without using temporal notions. The curious state of modern physics as outlined in Box 2 compounds the problem. Because quantum theories are obtained from classical theories by so-called quantization, and classical concepts are much closer to everyday experience, the language used by most physicists, myself included, often seems to imply that the classical theories are somehow deeper than the quantum theories obtained from them. But that is certainly only a reflection of our way to the truth. What is needed is a clear language in which to describe the quantum truth directly and an explanation, based on it, of why the world appears classical to us. I am proposing the notion of a Now as the basic quantum notion.

Getting to Grips with Elusive Time (p. 17) The idea that instants of time are distinct entities that should not be thought of as joined up in a linear sequence is a powerful intuitive experience for at least one non-scientist. A few days after the *Sunday Times* published its article 'Time's assassin' about my ideas in October 1998, I received by email a 'Question for Julian Barbour' from Gretchen Mills Kubasiak, who had read the article about me. She introduced herself with: 'I am merely a girl who lives in Chicago, works for a construction company and finds herself thoroughly captivated by your ideas. In fact, I have been unable to think of little else this past week.' She asked if she could put a question to me. Well, who could resist that request? I said yes, asking if by any chance, with her first name, she had German ancestry, and commented: 'I guess you know the German expression *Gretchenfrage* and its origin in Goethe's *Faust*, when Gretchen asks Faust about his attitude to religion and if he believed in God. It was especially nice to get your *Gretchenfrage.*' Subsequent correspondence persuades me that 'merely a girl' might not be the most accurate description of her, since she is a voracious reader and traveller (among much else). Some of her thoughts about time are worth passing on:

> Several weeks before I read the London *Times* article which brought your ideas to my attention, I started having a debate with a friend of mine on traveling. He stated that when a person travels between two places, it is the time spent on the

journey which makes the person able to appreciate and comprehend the final destination. Only by making a linear tour of the world and having a passage of time connect the two locations are we able to understand our final destination.

I disagreed. I have always believed that our lives are made up of individual moments that layer and co-exist with other moments, not a linear sequence of events. I did not accept his notion that time spent on a journey is relative to one's experience at their final destination. The passage of time, that for my friend constituted the journey, did not exist for me. That is not to say that what he viewed as his journey did not consist of moments but I could not accept that they were relative to the moment of the final destination simply because they preceded it.

Despite the fact that I had these beliefs in my head, I found that I lacked the vocabulary to make a satisfactory argument on paper. It is one thing to state your beliefs and quite another to be able to back up your argument. I had developed a few descriptive examples of moments in my life that I believed began to illustrate this idea but I knew of nothing that would support them.

One of my ideas addressed my moments with Buckingham Palace. As a small child I had listened to my mother recite the poem about Christopher Robin's visit to the changing of the guard and I stood silently alongside him and Alice. As a young girl I watched on television the newly married Prince and Princess of Wales venture forth onto the balcony to greet their public and I stood among the crowds. In both instances I was not 'there' and yet I was. When I actually stood in front of the palace as a teenager, the physical journey associated with that moment mattered not. What mattered were these other moments. When I stood in front of the palace, I was living not just that moment but co-existing with the other moments as well.

Then I came across the London *Times* article outlining your notion of the illusion of time and a spark of recognition within me was lit. Something I had always felt, but had never been able to express, was suddenly being put into words.

If, as you say, all moments are simultaneous and there is no linear sequence of events, does this not imply that the 'length' of a journey is completely irrelevant? If we exist in isolated moments, then the notion that time spent on a journey makes the experience cannot be true because time does not exist. If time is merely an illusion, the time spent on a journey is also an illusion.

My memories never fade. Memories from my supposed past shine as clearly as my present. I remember climbing out of my crib after a nap at 1 1/2 years old as clearly as I remember getting out of bed this morning. Aren't memories supposed to become less clear with time? These moments remain in my head as individual events. I rarely think of them in conjunction with moments that preceded or followed them. The memories in my head feel somewhat like a piece of sedimentary rock—as if these moments have all been compressed together and the connector pieces—the time that I thought held them together—has been blown away with the wind. These thoughts all exist simultaneously in my mind yet they reveal themselves to me one by one.

I think most important was my prevailing feeling of a stronger connection between moments perceived as being separated by time than between moments believed to be connected by time. What I am unclear about, however, is what causes this feeling of connection. Can there be a relationship between these moments? Not in the sense of a linear connection, but rather a feeling of empathy between them. To a certain extent, I think there is a subconscious awareness that there are these other moments occurring simultaneously and that there can be an acknowledgement between moments that are connected by subject matter.

If all moments are simultaneous, I am concurrently hearing the Christopher Robin poem being read, watching the Prince and Princess of Wales on the balcony, and standing in front of the Palace myself. My conscious mind feeds them to me in a linear sequence strung out with a bunch of other moments in an illusion of a continuous flow of action. While I am being read to, however, my subconscious is aware that I really am in front of Buckingham Palace and so a sense of really being there is brought to the Christopher Robin reading or to the Royal Wedding viewing.

This awareness that this other moment is occurring out there right now has struck me at many times. Sometimes it's when I'm reading a book, other times I'm walking down the street listening to music. Always, however, there is the feeling that I am somewhat connected to that other moment and I can almost feel there is the chance of stepping out of this moment and into another. It is the knowledge that there is another possibility to this moment.

To a certain extent, I often feel as if we are moving towards a timeless existence. The increasing usage of the computer by people on an everyday basis is one factor heading us in this direction. At any moment, without any thought to time, we can shop on our computers, chat, read newspapers, research, do our banking, etc. Also, more and more we are creating environments in which timelessness is the objectivity. Nowhere is this more obvious than in the twentieth-century environments of the department store, the amusement park and the casino. The goal is one dream-like moment, where there is no beginning and no end—no time.

Reading these comments again three months after they came, they strike me as often very close to my position. Incidentally, I address the original Gretchen's questions (*Glaubst du an Gott? Wie hält's du es mit der Religion?*) in the Epilogue.

Note for physicists (p. 18): Space plays two roles in Newtonian physics: it binds its contents together to form the plurality within the unity mentioned in this section (the separations between N objects in Euclidean space are constrained by both inequalities and algebraic relations, which give expression to this unity) and if defines positions at non-coincident times. In the type of physics I am advocating, only the first property is used, as will become clear in Part 3.

In relativity theory, the construction of 'three-dimensional' snapshots from two-dimensional photographs is greatly (but not insuperably) complicated by the fact that light travels at finite speed, so that objects are no longer where they seem to be. Readers familiar with relativity theory and concerned that my concept of a Now seems very non-relativistic are asked to defer judgment until Part 3. Einstein did not abolish Nows, he simply made them relative.

Laws and Initial Conditions (p. 22) Although Newton's and Einstein's laws work equally well in both time directions, there is one known phenomenon in quantum physics that seems to determine a direction of time at a truly fundamental level. It is observed in the decay of particles called kaons. Paul Davies discusses this phenomenon in some detail in his *About Time*. Most authors are agreed that this phenomenon does not seem capable of explaining the pronounced directionality of temporal processes, which is one of my main concerns in this book, but it is probably very important in other respects and may provide evidence that time really does exist as an autonomous governing factor in the universe. However, the evidence that it defines a direction in time is indirect, being based on something called the TCP theorem. Although this is most important in modern physics, what form if any it will take in the as yet non-existent theory of quantum gravity is not at all clear.

CHAPTER 2: TIME CAPSULES

The Physical World and Consciousness (1) (p. 26) There is a clear and detailed account of Boltzmann's ideas in Huw Price's book listed in Further Reading.

(2) (p. 27) It is worth quoting here two passages from Boltzmann himself. In 1895 he published (in perfect English—I wonder if he had assistance) a paper in *Nature* with the title 'On certain questions of the theory of gases'. It ends with a truly remarkable and concise statement of what much later became known as the *anthropic principle*. This expression was coined in 1970 by the English relativist Brandon Carter (who had earlier made important discoveries about the physics of black holes in the period leading up to Hawking's discovery that they can evaporate). The anthropic principle, which gained widespread attention initially through the book *The Anthropic Cosmological Principle* by John Barrow and Frank Tipler, expresses the idea that any universe in which intelligent life exists must have special and unexpected (from a purely statistical viewpoint) properties, since otherwise the intelligent life that observes these properties could not exist. Therefore we should not be surprised to find ourselves in a universe that does have special and remarkable properties.

In the following passage, the summits of the *H* curve to which Boltzmann refers correspond to states with very low entropy and high order. Note that Boltzmann credits his assistant with the idea.

> I will conclude this paper with an idea of my old assistant, Dr. Schuetz.
>
> We assume that the whole universe is, and rests for ever, in thermal equilibrium. The probability that one (only one) part of the universe is in a certain

state, is the smaller the further this state is from thermal equilibrium; but this probability is greater, the greater the universe itself. If we assume the universe great enough we can make the probability of one relatively small part being in any given state (however far from the state of thermal equilibrium), as great as we please. We can also make the probability great that, though the whole universe is in thermal equilibrium, our world is in its present state. It may be sayd [sic] that the world is so far from thermal equilibrium that we cannot imagine the improbability of such a state. But can we imagine, on the other side, how small a part of the whole universe this world is? Assuming the universe great enough, the probability that such a small part of it as our world should be in its present state, is no longer small.

If this assumption were correct, our world would return more and more to thermal equilibrium; but because the whole universe is so great, it might be probable that at some future time some other world might deviate as far from thermal equilibrium as our world does at present. Then the aforementioned H curve would form a representation of what takes place in the universe. The summits of the curve would represent the worlds where visible motion and life exist.

Boltzmann returned to this theme a year later, this time writing in German. The following is my translation:

One has a choice between two pictures. One can suppose that the complete universe is currently in a most unlikely state. However, one can also suppose that the eons during which improbable states occur are relatively short compared with all time, and the distance to Sirius is small compared with the scale of the universe. Then in the universe, which otherwise is everywhere in thermal equilibrium, i.e. is dead, one can find, here and there, relatively small regions on the scale of our stellar region (let us call them isolated worlds) that during the relatively short eons are far from equilibrium. What is more, there will be as many of these in which the probability of the state is increasing as decreasing. Thus, for the universe the two directions of time are indistinguishable, just as in space there is no up or down. But just as we, at a certain point on the surface of the Earth, regard the direction to the centre of the Earth as down, a living creature that at a certain time is present in one of these isolated worlds will regard the direction of time towards the more improbable state as different from the opposite direction (calling the former the past, or beginning, and the latter the future, or end). Therefore, in these small regions that become isolated from the universe the 'beginning' will always be in an improbable state.

Time Without Time (p. 29) In connection with my suggestion that the brain may be deceiving us when we see motion, it is interesting to note that, as Steven Pinker points out in his *How the Mind Works*, people with specific types of brain damage see no motion when normal people do see motion. In his words, they 'can see objects change their positions but cannot see them move—a syndrome that a philosopher once tried to convince me was logically impossible! The stream from a teapot does not flow but looks like an icicle; the cup does not gradually fill with tea but is empty and then suddenly full'.

If the mind can do these things, it may be creating the impression of motion in undamaged brains.

CHAPTER 3: **A TIMELESS WORLD**

First Outline (p. 36) The philosopher best known for questioning the existence of time and its flow was John McTaggart, who is often quoted for his espousal of the 'unreality' of time and the denial of transience. The following argument of his is very characteristic of professional philosophers:

> Past, present, and future are incompatible determinations. Every event must be one or the other, but no event can be more than one. If I say that any event is past, that implies that it is neither present nor future, and so with the others. And this exclusiveness is essential to change, and therefore to time. For the only change we can get is from future to present, and from present to past.
>
> The characteristics, therefore, are incompatible. But every event has them all. If [an event] is past, it has been present and future. If it is future, it will be present and past. If it is present, it has been future and will be past. Thus all the three characteristics belong to each event. How is this consistent with their being incompatible? (McTaggart 1927, Vol. 2, p. 20)

Some thoughts here certainly match my own thinking, especially that 'exclusiveness is essential to change', but McTaggart's arguments are purely logical and make no appeal to physics. Abner Shimony (1997)—to whom I am indebted for several discussions—compares McTaggart's position with mine, but I think he has not quite understood my notion of time capsules, so I do not feel that his arguments force me to accept transience.

A typical example of theological thought about time is this extract from *Conversations with God—An Uncommon Dialogue* by Neale Donald Walsch (kindly sent me by Ann Gill):

> Think of [time] as a spindle, representing the Eternal Moment of Now.
>
> Now picture leafs [sic] of paper on the spindle, one atop the other. These are the elements of time. Each element separate and distinct, yet each existing *simultaneously with the other*. All the paper on the spindle at once! As much as there will ever be—as much as there ever was . . .
>
> There is only One Moment—*this* moment—the Eternal Moment of Now (p. 29).

Again, there is some overlap with my position. Walsch's 'leafs', his elements of time, are my Nows. But the spindle of time, the Eternal Moment, is not at all part of my picture. My Nows are all constructed according to the same rule. There is no Eternal Moment, only the common rule of construction. I think Walsch is trying to grasp eternal substance where there is none, though I think he is right to say that the 'leafs' are all there at once and that this is a consoling thought. But we should not ask for more than we can get. Also, the image of time as a spindle is beautiful but misleading. In my view, the 'leafs' of time most definitely cannot be arranged along a single line, as the striking spindle image implies.

The Ultimate Arena (1) (p. 39) In this section I say that all structures that represent possible instants of time are three-dimensional. This is because the space we actually observe has three dimensions. However, in some modern theories (superstring theories) it is assumed that space actually has ten or even more dimensions. All but three of the dimensions are 'rolled up' so tightly that we cannot see them. In principle, my instants of time could fit into this picture. They would then have ten (or more) dimensions.

(2) This note is for experts. Platonia is a special type of configuration space known as a *stratified manifold*. The sheets, ribs and singular point that form the frontiers of Triangle Land are called *strata*. I believe that the stratified structure of Platonia is highly significant. Mathematicians and physicists really interested in this can consult DeWitt (1970) and Fischer (1970). The strata are generally regarded as something of a nuisance, since at them normal well-behaved mathematics breaks down. They are like grit in the works. But in the world's oyster they may be the grit from which grows 'a peal richer than all his tribe': not Desdemona, but time (Chapter 22). (After Othello had strangled Desdemona and then realized his dreadful mistake, he said before stabbing himself that he was 'one whose hand, Like the base Indian, threw away a pearl richer than all his tribe'.)

CHAPTER 4: **ALTERNATIVE FRAMEWORKS**

(1) (p. 61) I have written at considerable length about the early history of astronomy and mechanics and the absolute versus relative debate in my *Absolute or Relative Motion?* This has recently been reprinted as a paperback with the new title *The Discovery of Dynamics* (OUP, 2001). I still hope to complete a further volume bringing the story up to the present, and much has already been written, but my plans are in flux because of the developments mentioned at the end of the Preface and at various places in these notes. Readers wanting a full academic (and mathematical) treatment of the topics presented in Parts 2 and 3 of this book are asked to consult the above and the papers (Barbour 1994a, 1999, 2000, 2001), which cite earlier papers. For references to recent developments see p. 358 and my website (www.julianbarbour.com).

(2) (p. 64) In the main body of the text, I mention the importance of the fortunate circumstances of the world in enabling physicists to avoid worrying about foundations. Another very important factor is the clarity of the notion of empty space, developed so early by the Greek mathematicians, which deeply impressed Newton. He felt that he really could see space in his mind's eye, and regarded it as being rather like some infinite translucent block of glass. He and many other mathematicians pictured its points as being like tiny identical grains of sand that, close-packed, make up the block. But this is all rather ghostly and mysterious. Unlike glass and tiny grains of sand, which are just visible, space and its points are utterly invisible. This is a suspect, unreal world.

We are not bound to hang onto old notions. We can open our eyes to something new. Let me try to persuade you that points of space are not what mathematicians would sometimes have us believe. Imagine yourself in a magnificent mountain range, and that someone asked, 'Where are you?' Would you kneel down with a magnifying glass and look for that invisible 'point' at which you happen to be in the 'space' that the mountain range occupies? You would look in vain. Indeed, you would never do such a silly thing. You would just look around you at the mountains. *They tell you where you are.* The point you occupy in the world is defined by what the world looks like as seen by you: it is a snapshot of the world as seen by you. Real points of space are not tiny grains of sand, they are actual pictures. To see the point where you are in the world, you must look not inward but *outward.*

The plaque near the grave of Christopher Wren in St. Paul's Cathedral says simply: 'If you seek a monument, look around you.' The point where you are is a monument too, and you see it by looking around you. It is this sort of change of mindset that I think we need if we are to understand the universe and time.

To conclude this note, a word about what is perhaps the most serious problem in my approach. It is how to deal with infinity. As so far defined, each place in Platonia corresponds to a configuration of a finite number of objects. Such a universe is like an island of finite extent. One could allow the configurations to have infinite extent and contain infinitely many objects. That is not an insuperable problem. The difficulty arises with the operations that one needs to perform. As presented in this book, the operations work only if the points in Platonia, the instants of time, are in some sense finite. There may be ways around this problem—Einstein's theory can deal beautifully with either finite or infinite universes—but infinity is always rather difficult. There is something 'beyond the horizon', and we can never close the circle of cause and effect. In short, we cannot build a model of a completely rational world. Precisely for this reason Einstein's first and most famous cosmological model was spatially finite, closed up on itself. The constructions of this book are to be seen as a similar attempt to create a rational model of the universe in which the elusive circle does close.

In fact, if the work with Niall Ó Murchadha mentioned at the end of the Preface, which suggests that absolute distance can be eliminated as a basic concept (see Box 3), can be transformed into a complete theory, the problem of infinity may well be solved in the process. If size has no meaning, the distinction between a spatially finite or infinite universe becomes meaningless.

CHAPTER 5: NEWTON'S EVIDENCE

The Aims of Machian Mechanics (1) (p. 71) In creating the beautiful diagrams that form such an important part of this section, Dierck Liebscher was able to draw on initial data devised by Douglas Heggie (University of Edinburgh), using software written by Piet Hut (Institute for Advanced Study), Steve McMillan (Drexel University) and Jun Makino (University of Tokyo). Dierck has written a very interesting book (alas, as yet published only in German) on the connection between different possible geometries and Einstein's relativity theory (Liebscher 1999). It contains many striking computer-generated diagrams.

(2) Poincaré's discussion is contained in his *Science and Hypothesis,* which, along with the writings of his contemporary, Mach, became a popular-science best-seller. In fact, in this book I am actually revisiting many of the themes discussed by Poincaré and Mach, but with the advantage of hindsight. How are the great issues they raised changed by the discovery of general relativity and quantum mechanics? I have adapted Poincaré's discussion somewhat to match the requirements of a timeless theory (he considered only the possibility of eliminating absolute space).

(3) Since writing Box 3, which draws attention to the present unsatisfactory use of absolute distance in physics, I have discovered a way to create dynamical theories in which distance is not absolute. This is achieved by a very natural extension of the best-matching idea described later in the book. The new insights that I mention in the Preface are in part connected with this development. One of the most exciting is that, if such theories do indeed describe the world, gravitation and the other forces of nature are precisely the mechanism by means of which absolute distance is made irrelevant. Since this work is still in progress, I shall make no attempt to describe it in detail, but I shall keep my website (www.julianbarbour.com) up to date with any progress (see also p.358).

CHAPTER 6: THE TWO GREAT CLOCKS IN THE SKY

The Inertial Clock (p. 99) Tait's work, which I feel is very important, passed almost completely unnoticed. This is probably because two years later the young German Ludwig Lange introduced an alternative construction for finding inertial frames of reference, coining the expression 'inertial system'. Lange deserves great credit for bringing to the fore the issue of the determination of such systems from purely relative data, but Tait's construction is far more illuminating. Lange's work is discussed in detail in Barbour (1989) and Tait's in Barbour (forthcoming).

The Second Great Clock (p. 107) A very nice account of the history of the introduction of ephemeris time was given by the American astronomer Gerald Clemence (1957).

CHAPTER 7: **PATHS IN PLATONIA**

Nature and Exploration (p. 109) For physicists and mathematicians who do not know the book, a wonderful account of the variational principles of mechanics, together with much historical material, is given by Lanczos (1986).

Developing Machian Ideas (p. 115) Translations of the papers by Hofmann, Reissner and Schrödinger, along with other historical and technical papers on Mach's principle, can be found in Barbour and Pfister (1995).

Exploring Platonia (p. 115) The special properties of Newtonian motions with vanishing angular momentum were discovered independently of the work of Bertotti and myself by A. Guichardet in the theory of molecular motions and by A. Shapere and F. Wilczek in the theory of how micro-organisms swim in viscous fluids! A rich mathematical theory has meanwhile developed, and is excellently reviewed in the article by Littlejohn and Reinsch (1997), which contains references to the original work mentioned above. All mathematical details, as well as references to the earlier work by Bertotti and myself, can be found in Barbour (1994a).

CHAPTER 8: **THE BOLT FROM THE BLUE**

Historical accidents (p. 123) Poincaré's paper can be found in his *The Value of Science*, Chapter 2. Pais's book is in the Bibliography.

Background to the Crisis (p. 124) The best (moderately technical) historical background to the relativity revolution that I know of is the book by Max Born. It is available in paperback.

The Forgotten Aspects of Time (p. 135) My claims about the topics that somehow escaped Einstein's attention are spelled out in detail in Barbour (1999, forthcoming). I have tried to make good the gap in the literature on the theory of clocks and duration in Barbour (1994a).

CHAPTER 10: **THE DISCOVERY OF GENERAL RELATIVITY**

Einstein's Way to General Relativity (p. 151) Einstein's papers and correspondence are currently being published (with translations into English) by Princeton University Press. The letter to his wife mentioned in this section can be found in the first volume of correspondence (Stachel *et al.* 1987).

CHAPTER 11: **GENERAL RELATIVITY: THE TIMELESS PICTURE**

Platonia for Relativity (p. 167) This is a technical note about the definition of superspace. The equations of general relativity lead to a great variety of different kinds of solution, including ones in which there are so-called closed time-like loops. These are solutions in which a kind of time travel seems to be possible. The question then arises of whether a given solution of general relativity—that is, a space-time that satisfies Einstein's equations—can be represented as a path in superspace, in technical terms, as a unique succession of Riemannian three-geometries. If this is always so, then superspace does indeed seem a natural and appropriate concept. Unfortunately, it is definitely not so. There are two ways in which we can attempt to get round this difficulty. We could say that classical general relativity is not the fundamental theory of the universe, since it is not a quantum theory. This allows us to argue that superspace is the appropriate quantum concept and that it will allow only certain 'well-behaved' solutions of general relativity to emerge as approximate classical histories. For these, superspace will be an appropriate concept. Alternatively, we could extend the definition of superspace to include not only proper Riemannian 3-geometries (in which the geometry in small regions is always Euclidean), but also pseudo-Riemannian 3-geometries (in which the local geometry has a Minkowski type signature), and also geometries in which the signature changes within the space. For the reasons given in the long note starting on p. 348 below, I prefer the second option.

The above note was written before my new insights mentioned at the end of the Preface. I now believe that there is a potentially much more attractive resolution of the difficulty: the true arena of the world is not superspace but conformal superspace, which I describe on p. 350.

Catching Up with Einstein (1) (p. 175) Figure 30 is modelled directly on well-known diagrams in Wheeler (1964) and Misner et al. (1973).

(2) Technical note: Einstein's field equations relate a four-dimensional tensor formed from geometrical quantities to the four-dimensional energy–momentum tensor, which is formed from the variables that describe the matter. Machian geometrodynamics shows how these four-dimensional tensors are built up from three-dimensional quantities. The two principles by which this is done are best matching, and Minkowski's rule that the space and time directions must be treated in exactly the same way (see the following note). As far as I know, the mathematics of how this is done when matter is present was first spelled out in a recent paper by Domenico Giulini (1999), to whom I am indebted for numerous discussions on this and many other topics covered in this book.

A Summary and the Dilemma (1) (p. 177) This is another technical note. My image of space-time as a tapestry of interwoven lovers rests on the following property of Einstein's field equations. If, in any given space-time that is a solution of the field equations, we lay out an arbitrary four-dimensional grid in any small region of the space-time, we can then, in principle, attempt to take the data on one three-dimensional hypersurface and use Einstein's equations to evolve these data and recover the space-time in the complete region. Normally, we attempt to do this in a time-like direction. However, the *form* of the equation is exactly the same whichever direction in which we choose to attempt the evolution from initial data. This is an immediate consequence of an aspect of the relativity principle that Minkowski gave a special emphasis: as regards the structure of the equations, whatever holds for space holds for time and vice versa.

What is more, however we choose the 'direction of attempted evolution', Einstein's equations always have a very characteristic structure. There are ten equations in all. One of them does not contain any derivative with respect to the variable in which we are going to attempt the evolution. Three of them contain only first derivatives with respect to that variable. The remaining six equations contain second derivatives with respect to it and have the form of equations that are suitable for evolution in the chosen direction. But we must first solve the other four equations, which are so-called constraints. Unless the initial data satisfy these four equations, evolution is impossible.

There are two ways to look at a space-time that satisfies Einstein's equations: either as a structure obtained from initial data that have been (somehow) obtained in a form that satisfies the constraints and then built by the more or less conventional evolution equations, or as a structure that satisfies everywhere the constraints however we choose to draw the coordinate lines. In the second way of looking at space-time, conventional evolution does not come into the picture at all. Much suggests that this is the more fundamental way of looking at Einstein's equations (see, in particular, Kuchař's beautiful 1992 paper). The connection with my timeless way of thinking about general relativity is expressed by the fact that the three constraint equations containing only first derivatives of the evolution variable are precisely the expression of the fact that a best-matching condition holds along the corresponding 'initial' hypersurface, while the fourth constraint equation, containing no derivatives of the evolution variable, expresses the fact that proper time is determined in geometrodynamics as a local analogue of the astronomers' ephemeris time. It is this complete freedom to draw coordinate lines as we wish and, at least formally, to attempt evolution in any direction, that makes me feel that the second alternative envisaged in the Platonia for Relativity note is appropriate. I think it is also very significant that Einstein's equations have the same form whatever the signature of space-time. The signature is not part of the equations, it is a condition normally imposed on the solutions. The demonstration that Einstein's general relativity is the unique theory that satisfies the criterion (mentioned at the end of this section) of a higher four-dimensional symmetry was given by Hojman *et al.* (1976).

I mentioned on p. 346 at the end of the notes on Chapter 4 my recent discovery of a way to create dynamical theories of the universe in which absolute distance is no longer relevant. My Irish colleague Niall Ó Murchadha, of University College Cork, and I are currently working on the application of the new idea to theories like general relativity, in which geometry is dynamical. There is a possibility that this work will not only give new insight into the structure of general relativity, in which a kind of residual absolute distance does play a role, but also lead to a rival alternative theory in which no distance of any kind occurs.

The key step is to extend the principle of best matching from superspace to so-called conformal superspace. In the context of geometrodynamics, this is analogous to the passage from Triangle Land to Shape Space as described in Box 3. However, whereas in Box 3 it is only the overall scale that is removed, and it is still meaningful to talk about the ratios of lengths of sides, the transition to conformal superspace is much more drastic and removes from physics all trace of distance comparison at spatially separated points.

In more technical terms, for people in the know, each point of conformal superspace has a given conformal geometry and is represented by the equivalence class of metrics related by position-dependent scale transformations.

The potentially most interesting implication of this work is that it could resolve the severe problem of the criss-cross fabric of space-time illustrated by Figure 31. At the level of conformal superspace, the universe passes through a unique sequence of states. For latest developments, please consult my website (www.julianbarbour.com) and the final entries in these notes and the notes on p.358.

CHAPTER 12: THE DISCOVERY OF QUANTUM MECHANICS

(p. 191) On the connection between particles and fields, let me mention here that I assume the appropriate 'Platonic' representation at the level of quantum field theory to be in terms of the states of fields, not particles.

CHAPTER 13: THE LESSER MYSTERIES

(p. 202) Wheeler and Zurek (1983) have published an excellent collection of original papers on the interpretational problems of quantum mechanics.

CHAPTER 14: THE GREATER MYSTERIES

The Many-Worlds Interpretation (p. 221) Everett's original Ph.D. thesis, his published paper and the papers of DeWitt (and some other people) relating to the many-worlds idea can be found in the book by DeWitt and Graham (1973).

CHAPTER 16: 'THAT DAMNED EQUATION'

History and Quantum Cosmology (p. 240) More details on the Leibnizian idea that the actual universe is more varied than any other conceivable universe are given in Smolin (1991), Barbour and Smolin (1992), and Barbour (1994b). The quotation from T. S. Eliot is in Eliot (1964). My book is Barbour (1989).

'That Damned Equation' (p. 247) Technical note: In connection with Chapter 11, it is interesting that the form of the Wheeler-DeWitt equation is independent of the signature of space-time.

(1) (p. 247) For physicists I should mention that there is an important alternative to regarding the Wheeler–DeWitt equation as analogous to the stationary Schrödinger equation. It also bears a resemblance to the relativistic Klein–Gordon equation, the role of time in that equation being played, essentially, by the volume of the universe in the case of the Wheeler–DeWitt equation.

(2) (p. 247) Kuchař's objections to my timeless interpretation of the Wheeler–DeWitt equation can be found in the discussion sessions at the end of Barbour and Pfister (1995). Comprehensive reviews of the problems of time in quantum gravity can be found in Kuchař (1992) and Isham (1993).

(3) (p. 247) In discussions with me in 1994 at an international conference on quantum gravity held at Durham, Bryce DeWitt expressed two main reservations about his 'damned equation'. The first was that it required a division of space-time into space and time, which he felt was running counter to the great tradition of relativity initiated by Einstein and Minkowski. I have already explained why I feel that this may not necessarily be an objection; indeed, it may not be possible to give objective content to general relativity unless such a split is made. DeWitt's second objection was that the 'damned equation' had not as yet yielded any concrete results and was (is) plagued with mathematical difficulties. This is certainly true, and I have omitted all discussion of these difficulties, which are certainly great. However, I think it is worth noting that as physicists' understanding of the equations that describe nature becomes deeper, the equations themselves become more sophisticated and harder to solve. It is much harder to find solutions of Einstein's equations than Newton's. This tendency—deeper understanding of principles bringing with it greater intractability of equations—will almost certainly mean that progress in quantum gravity is very slow. In fact, for over a decade, a group centred on the relativist Abhay Ashtekar, including my friend Lee Smolin and another friend Carlo Rovelli, has been working intensively on a particular approach to canonical quantum gravity (the broad framework in which DeWitt derived his equation) and have certainly resolved some of the difficulties. An account of this work can be found in Lee's *The Life of the Cosmos* and *Three Roads to Quantum Gravity*. Kuchař too has made many important contributions.

If the ideas described in the note on p. 350 work out as Niall Ó Murchadha and I believe they could, the difficult issues raised in the final part of Chapter 16 and in the above notes will be to a very large degree resolved. The conceptual uncertainties about the correct way to proceed that have plagued the theory for four decades could all be removed. Both for general relativity and the alternative theory that might replace it, the wave function of the universe will certainly be static and give probabilities for configurations as explained in the main text. The main difference is that only the intrinsic structure will count, so that all configurations that have the same structure and differ only in the local scales will have the same probability. They will merely be different representations of the same instants of time. However, for general relativity there will be a curious residual scale that represents a volume of the universe. It will be meaningful to say that the universe has a volume but not how the volume is distributed between the intrinsic structures contained within in.

CHAPTER 17: **THE PHILOSOPHY OF TIMELESSNESS**

(p. 255) On the subject of the aims and methods of science, I strongly recommend David Deutsch's *The Fabric of Reality.*

CHAPTER 18: **STATIC DYNAMICS AND TIME CAPSULES**

Dynamics Without Dynamics (p. 258) In this section I refer to investigations by various authors. Their studies will be found in the bibliography. Physicists really interested in the semiclassical approach may also like to consult the review article by Vilenkin (1989), the paper by Brout (1987), the final part of Zeh (1992, 1999) and the introductory article by Kiefer (1997). The fullest account of my own ideas is Barbour (1994a).

CHAPTER 19: **LATENT HISTORIES AND WAVE POCKETS**

Schrödinger's Heroic Failure (p. 278) In the first draft of this book I included a long section on the very interesting interpretation of quantum mechanics advanced originally by de Broglie, and revived by Bohm, whose 1952 paper I strongly recommend to physicists together with Peter Holland's book (Holland 1993). With regret I omitted it, as I felt that it made this book too long, especially since I believe that the interpretation does not really solve the problem. However, I particularly value the way in which it shows that all the results of quantum mechanics can be obtained in a framework in which positions are taken as basic. This made the theory attractive to John Bell, as we shall see in the next chapters.

CHAPTER 20: **THE CREATION OF RECORDS**

The Creation of Records: First Mechanism (1) (p. 284) Bell's paper can be found in his collected publications *Speakable and Unspeakable in Quantum Mechanics.*

(2) (p. 284) Mott's paper is reproduced in Wheeler and Zurek (1983). Heisenberg's treatment is in his *Physical Principles of Quantum Theory.* I am very grateful to Jim

Hartle, who first drew my attention to Mott's paper. At that time he was considering seriously an interpretation of quantum cosmology that is quite close to my own present position. He has since backed away somewhat, and now advocates an interpretation of quantum mechanics in which history is the fundamental concept. I should also like to express my thanks here to Dieter Zeh. Zeh, who was in this business long before me, also made me realize the importance of Mott's work, and, crucially, alerted me to Bell's paper. There are not many physicists who take the challenge of timelessness utterly seriously, but Dieter Zeh and his student Claus Kiefer, from both of whom I have gained and learned much, are two of them.

CHAPTER 21: THE MANY-INSTANTS INTERPRETATION

Bell's 'Many-Worlds' Interpretation (p. 299) In his 'cosmological interpretation' of quantum mechanics, Bell combined elements derived from both Everett and the de Broglie–Bohm interpretation (see the note to Chapter 19). In fact, Bell's account of his mixed interpretation is rather terse, and can be misunderstood. I am most grateful to Fay Dowker and Harvey Brown for drawing my attention to an error I made in reporting Bell's idea in my first draft of this book. In this section I follow their interpretation of Bell, which I am sure is what he did mean.

The Many-Instants Interpretation (p. 302) I hope I have made it clear that probability 'to be experienced' or 'to exist' is a problematic concept. If consciousness is determined by structure, the consciousness is already in the Nows and must be experienced irrespective of their probabilities. What role remains for probabilities? It is a very difficult issue. Probability is already puzzling in ordinary quantum mechanics, and even in classical physics. Cold water could boil spontaneously, but we never see this happen. Standard probability arguments suggest that what is possible but hugely improbable will not be experienced. Much suggests that probabilities in some form are inescapable in quantum theory simply because it explores mutually exclusive possibilities. Instants of time are natural candidates for the ultimate exclusive possibilities. If certain very specially structured instants do get hugely larger probabilities than others, and are the ones habitually experienced, that must, I feel, count as explanation. But as an indication of the depth of this problem, I add here in Box 16 an edited email exchange I had with Fay Dowker of Imperial College, London. I had especially asked her to read my first draft, since she is a very clear thinker but is sceptical about both many worlds and canonical quantization, the approach to quantum gravity that I favour.

CHAPTER 22: THE EMERGENCE OF TIME AND ITS ARROW

Soccer in the Matterhorn (p. 307) In the second edition of *The Physical Basis of the Direction of Time*, Zeh says that the intrinsic dynamical asymmetry of quantum gravity 'offers the possibility of *deriving* an arrow of time (perhaps even without imposing any special conditions)'.

Timeless Descriptions of Dynamics (1) (p. 309) For specialists: for each stage of a perturbation expansion, Mott always chooses a kernel in an integral representation corresponding to outgoing waves. However, nothing in the mathematics rules out the (occasional) choice of incoming waves. This would mess up everything.

(2) (p. 309) I should emphasize that Mott, like Bell, never used any expression like 'time capsule', and clearly did not think in such terms about alpha-particle tracks. Neither did Mott's work on alpha-particle tracks seem to have prompted him to any intimations of a many-worlds type interpretation of quantum mechanics. I learned this from Jim Hartle. Over a decade ago, when collaborating with Stephen Hawking in Cambridge, Jim lodged at his college, Gronville and Caius (featured famously in *Chariots of Fire*), which was also Mott's college. Over dinner Jim asked Mott whether his paper had not led him to anticipate some form of Everett's idea, and was told no. Apparently, all the 'young Turks' followed the Copenhagen line without hesitation at that time. Shortly before his death about two years ago, when he was still mentally very alert, I contact Mott and asked if I could talk to him about his paper. Alas, he was too ill to keep the appointment, telling his secretary he was very disappointed 'since the man wanted to talk about work I did nearly sixty years ago'.

A Well-Ordered Cosmos? (p. 321) This final section follows closely the final section of Barbour (1994a).

BOX 16 An Email Dialogue

DOWKER. It seems to me that you provide no scheme for making predictions, and I would further claim that no such scheme can exist which contains the two aspects that are fundamental to your scheme: canonical quantum gravity (CQG) and the Bell version of the many-worlds interpretation (MWI).

BARBOUR. I think you are right, subject to what one means by prediction. I cannot make the kinds of prediction you want, and you correctly identify the reasons. I feel the arguments for CQG and MWI outweigh desire for predictions of the kind you would like.

DOWKER. I freely admit that I am rather attached to the notion of the universe (and I) having had, and being about to have, a continuous history. But my criticism here is not the absence of history in your approach, but, to repeat, that there is no way to make predictions about the results of our observations. In my view this is a deficiency that cannot be overcome. Whatever else science tells us about the world, it must allow us to make predictions about our observations that we can check.

BARBOUR. I am not sure we can impose such a criterion on Nature. The Greeks had the notion of saving appearances (finding a rational explanation for the phenomena we observe) that is already very valuable. You may be asking more of Nature than she is prepared to give.

DOWKER. In backing up this criticism I shall focus on the aspect that I am most familiar with and on which I have most confidence in my own views, which is the aspect of the interpretation of quantum mechanics. I am pleased that your book draws attention to the work of Bell on the many-worlds interpretation, since it has not had the recognition it deserves. In my view his version is the only well-defined many-worlds interpretation (I'll call it BMWI) that exists in the literature.

BARBOUR. I agree it is well defined, but with reservations about the role of time. The time of an observation, like any other observable, must be extracted from present records. When you start to ask how that is done in practice and how Nature does put time into the records, I think things may become less well defined. I do believe that almost all physicists this century have blindly followed Einstein in declining to try to understand duration at a fundamental level. A lot of the first part of my book is about that. I think my position might be stronger than you realize.

DOWKER. I think that neither your version (which I'll call JMWI) nor BMWI allows us to make predictions about what we observe (so I disagree with Everett's statement 'the theory itself predicts that our experience will be what it in fact is'). Let me take your version. There we have many configurations at time t. The most serious problem is that in a scheme like yours, in which all the possibilities are realized, there is no role for the probabilities. The usual probabilistic Copenhagen predictions for the results of our observations cannot be recovered. An excellent reference which analyses the MWI literature and the various attempts to derive the Born interpretation from MWI is Adrian Kent [1990, *International Journal of Modern Physics,* **A5**, 1745]. Adrian concludes that they fail. I'll just state again the main reason that they fail: when *all* the elements in a sample space of possibilities are realized, then probability is not involved. Your idea is that it is the sample space itself, i.e. how many copies of each configuration are included in the sample space, which is determined by the (squares of the) coefficients of the terms in the wave function. That is all well and good (if bizarre). But there's no reason then to call those numbers probabilities, and no way to recover the probabilistic predictions of Copenhagen quantum mechanics. In fact the MWI proponents themselves agree that the failure to reproduce the Copenhagen predictions is a problem and do try to address it, but without success.

BARBOUR. I accept that this is a strong critique. I nevertheless feel that my scheme does in principle have predictive strength. If you could see Platonia and Born's probability density concentrated incredibly strongly on a tiny proportion of its points that all turn out to be time capsules as I define them and Bell describes them, would you not find that impressive, and something like a rational explanation for our experiences?

DOWKER. As well as the MWI, you base your conjecture of timelessness on the technical result that when a canonical quantization scheme is applied to general relativity, the wave function cannot contain the time. My understanding of the state of affairs in canonical quantum gravity is that, because of this, no one knows how to make the kind of predictions we'd like to make: explanations such as 'What happens in the final stages of black hole collapse?', 'Why is the cosmological constant so very small?', etc.

BARBOUR. I agree with your first example (and do not think it is too serious—there may be questions that it is just not sensible to ask), but in principle my scheme could predict that virtually all time capsules will appear to have been created in nearly classical universes with a very small cosmological constant. After all, that is what our present records indicate. If all probable configurations seem to contain records that indicate a small cosmological constant, I am okay.

DOWKER. My reaction to the situation is that formulating general relativity in a canonical way has been shown to be the wrong thing to do—we did what we weren't supposed to—divided up space-time into space and time again. Even if it wasn't clear from the beginning that it would be incredibly difficult to maintain general covariance of the theory whilst trying to treat space and time differently, I find the lack of any insight into how to recover predictions within the canonical quantum gravity program convinces me that we should look elsewhere for a quantum theory of gravity.

BARBOUR. As he was creating general relativity, Einstein was convinced general covariance had deep physical significance. Two years later, correctly in my opinion, he completely abandoned that position. In my opinion, general covariance is an empty shell (I say something about this at the end of Chapter 10 and in the notes to it). I believe it is not possible to give any meaning to the objective content of general relativity without saying how the three-dimensional slices in space-time are related to each other. That is the very content of the theory. That is why I think the arguments for canonical quantum gravity are very strong indeed. The constraints of the canonical theory are its complete content.

DOWKER. Having made my basic points, let me now just say that I find it incredibly hard to understand how, as a solipsist of the moment, you must view science and the scientific enterprise.

BARBOUR. Answered above I think. Science should explain what we observe. We habitually observe and experience time capsules. Even granting the real difficulties with calling the square of a static amplitude a probability, should it turn out that the Wheeler–DeWitt equation does strongly concentrate the square of the amplitude on time capsules, I think that would be an incredibly strong and suggestive result.

DOWKER. Take the idea that a good scientific theory should be falsifiable.

BARBOUR. I think my idea is falsifiable in the following sense. There may well be configurations of the universe with records of my idea and mathematical proofs that the Wheeler–DeWitt equation most definitely does not concentrate the square of the amplitude on time capsules. If I too am in them, I would have to say my proposal for an explanation of why we think time flows has failed.

DOWKER. That presupposes that there will be a future in which we can try new experiments that test the theory and find that these experiments may be in contradiction to our predictions. The very word 'prediction', which I have used so many times in this letter, is laden with time-meaning. A prediction is a statement of expectation of something that will happen. Prediction is the lifeblood of science. How could we do science without it?

BARBOUR. I totally agree about the importance of prediction. But it does not necessarily have to involve time in the way you suggest. From observations of one side of the Moon, astronomers tried to predict what was on the other side. They got it wrong when the other side was seen. I do not think time comes into such predictions at all significantly. Consider, as Jim Hartle once did when he was quite close to my present position, geology. The rocks of the Earth hardly change. Suppose the idea of continental drift had been proposed before America had been discovered. It would have predicted the existence of America and the geology of its east coast (the west of Ireland exactly matches Newfoundland, I believe). Again, time is not essentially involved in this prediction. I think Bell puts my case very well: 'We have no access to the past. We have only our "memories" and "records". But these memories and records are in fact *present* phenomena.' The italics are Bell's. Predictions are always verified in the present. That is my apologia.

Notes Added for This Printing.

As mentioned at the end of the Preface and at various places in the Notes, there have been some promising developments of the ideas presented in this book since it was sent to press in spring 1999. They are contained in two joint papers published electronically and available on the web: Julian Barbour and Niall Ó Murchadha, 'Classical and quantum gravity on conformal superspace', http//xxx.lanl.gov/abs/gr-qc/9911071 and Julian Barbour, Brendan Z. Foster, and Niall Ó Murchadha, 'Relativity without relativity', http//xxx.lanl.gov/abs/gr-qc/0012089 [the xxx is correct].

The potential significance of the first paper has already been explained on p. 349/350. At this stage, I do not wish to make any firm statements about this new work since it is incomplete and has not yet been exposed to scrutiny by other physicists, but I can at least give some idea of what is at stake. The basic issue is the status of the relativity principle. When Einstein and Minkowski created special relativity, they deliberately made no attempt to explain the remarkable structure that their work had brought to light: the existence of spacetime and its associated light cone, both being reflected in the Lorentz invariance of the laws of nature. They adopted Lorentz invariance, which assumes the existence of length, as the basis of physics. In small regions of spacetime, this still remains true in general relativity.

Taken together the two papers cited above suggest that all of the presently known facts of relativity and electromagnetism can be derived in a new and hitherto unsuspected manner from three assumptions: 1) an independent time plays no role in dynamics; 2) best matching (pp. 116/7) is the essential element in the action principle of the universe; 3) any theory satisfying these principles must have nontrivial solutions. It is the third assumption that makes a dramatic difference. Hitherto, in common with other colleagues, I had assumed (see p. 181) many different theories could satisfy the first two conditions, but, as my collaborator Niall Ó Murchadha discovered, this is not the case. The reasons for this and its remarkable potential consequences are spelled out in the second cited paper. It is frustrating not to be able to say more at the present moment, but at a time of uncertainty about the final outcome it is better to say less rather than too much. My website (julianbarbour.com) will carry more detailed information.

I conclude with some additional comments that have mostly already appeared in the UK paperback.

My website now carries an email dialogue between myself and Don Page (whom Hawking, in *A Brief History of Time*, credits with pointing out, with Raymond Laflamme, 'his greatest blunder'). I think Don has made some interesting and valuable comments, including both valid criticism of some points but also reassurance on other issues on which I had doubts. Don is another person who takes the timelessness of physics utterly seriously, and, in fact, our views are very close. He has written several papers on the problem of consciousness and quantum mechanics (Sensible Quantum Mechanics). Full details can be found on my website. I should also like to mention here an idea that Dieter Zeh put forward in the meeting at Huelva in Spain at which I conducted my straw poll. This is that the universe must, of necessity, always be observed as expanding. I find this an intriguing idea and, if it is correct, it would fit beautifully with the open-ended, flowerlike structure of Platonia. I think Dieter's idea, which I hope is correct though I have hedged my bets in this book, influenced me somewhat in the writing of Box 3 and parts of the Epilogue.

I regret that in the final chapter I made no mention of the idea of inflation, which is explained in a gripping and candid book by Alan Guth that I have at least now included in the books recommended for further reading. From reading Guth's book I also learned that an interesting proposal of a mechanism for the 'creation of the universe' was made by Edward Tryon in 1973. I should also have mentioned that in 1982 Alexander Vilenkin proposed an influential alternative proposal to Hawking's no-boundary idea (1981) and that Jim Hartle played a significant role in its development, which culminated the Hartle-Hawking wave function. My apologies to these authors (none of whom have registered any complaint). Details can be found in Guth's book.

It has also been pointed out to me by several email correspondents that there is a clear anticipation of some of my ideas about time in Fred Hoyle's novel *October the First Is Too Late*, which Paul Davies discusses in his *About Time*. Sir Fred's 'pigeonholes' are essentially my time capsules. American reviewers and correspondents also noted a similarity with the philosophy of time that underlies Kurt Vonnegut's Slaughterhouse Five. Having now read the book, I can confirm that this is the case. Of course, as I make clear in the text, John Bell also formulated the idea of time capsules (without giving them any name) quite clearly long before me. Another correspondent, Andrew Clifton, regretted that I had not devoted at least a few words to demolishing the idea that there really is a 'moving present'. I think he was right. Happily, David Deutsch has done the job very well in his The *Fabric of Reality*.

I should also like to thank Damien Broderick, who has reviewed my book for The Australian, for drawing my attention to various misprints.

It is also now clear to me that in the body of the text I should have said more about possible ways in which my ideas could be refuted. A theory is no use to science unless it is capable of disproof. In the email exchange with Fay Dowker, I did mention the possibility of mathematical disproof of my conjecture that the Wheeler—DeWitt equation concentrates its solutions on time capsules. However, I think that (in normal parlance) that might take decades. Something that might occur much sooner is a completely convincing definitive form of superstring theory (or some other unified theory) that reintroduces an external time (string theory does currently use background structures). That would kill my idea. My own feeling is that in fact superstring theory will, if and when it is found, turn out to be timeless.

Then there is one other quite different way in which my ideas could be disproved. This is if experimental evidence can be found that shows collapse of the quantum wave function to be a real physical process. In this connection, I should like to mention especially an experiment proposed by Roger Penrose to test this very possibility. He is developing it in collaboration with Anton Zeilinger, the Austrian physicist based in Vienna, who has performed so many incredibly beautiful quantum experiments. Penrose, very understandably, finds the many-worlds interpretation of quantum mechanics extremely hard to accept (see my comments about the death of Diana), and with great persistence is trying to find a way round it. He has certainly identified the greatest single issue in modern physics. If his experiment, which could perhaps be performed within a decade, works out in the way he hopes, it will be a huge development and destroy my approach (because it will show that quantum mechanics does not hold macroscopically). The volume containing my paper (Barbour 2000) also contains Penrose's most important paper on the subject, and also a related paper by Joy Christian, who was a student of Abner Shimony. Joy, following Abner (see my Epilogue), is trying to establish transience as a real physical thing. I think that if collapse of the wave function could be demonstrated as a real physical phenomenon, that would be true demonstration of something that one might call transience.

JB
January 2001

FURTHER READING

Barrow, John, 1992, *Theories of Everything,* Oxford University Press, Oxford.

Barrow, John and Tipler, Frank, 1986, *The Anthropic Cosmological Anthropic Principle,* Clarendon Press, Oxford.

Bondi, Hermann, 1962, *Relativity and Common Sense: A New Approach to Einstein,* Dover, New York.

Coleman, James, 1954, *Relativity for the Layman,* William-Frederick Press, New York (1990, Penguin, London).

Coveney, Peter and Highfield, Roger, 1991, *The Arrow of Time,* Flamingo, London.

Davies, Paul, 1991, *The Mind of God,* Simon & Schuster, New York.

Davies, Paul, 1995, *About Time: Einstein's Unfinished Revolution,* Penguin Press, London.

Deutsch, David, 1997, *The Fabric of Reality,* Penguin Press, London.

Eddington, Arthur, 1920, *Space, Time and Gravitation,* Cambridge University Press, Cambridge.

Einstein, Albert, 1960, *Relativity: The Special and the General Theory. A Popular Exposition,* Routledge, London.

Greene, Brian, 1999, *The Elegant Universe: Superstrings, Hidden Dimensions, and the Quest for the Ultimate Theory,* Vintage Books, New York.

Gribbin, John, 1984, *In Search of Schrödinger's Cat: Quantum Physics and Reality,* Corgi, London.

Guth, Alan H., 1997, *The Inflationary Universe: The Quest for a New Theory of Cosmic Origins,* Perseus Books Group, New York.

Lippincott, Kristen, Eco, Umberto, Gombrich, E. H. *et al.*, 1999, *The Story of Time,* Merrell Holberton, London . [This is a splendid book on the most diverse aspects of time in culture and science.]

Lockwood, Michael, 1989, *Mind, Brain and the Quantum,* Basil Blackwell, Oxford.

Novikov, Igor, 1998, *The River of Time,* Cambridge University Press, Cambridge.

Penrose, Roger, 1989, *The Emperor's New Mind: Concerning Computers, Minds, and the Laws of Physics,* Oxford University Press, Oxford.

Price, Huw, 1996, *Time's Arrow and Archimedes' Point,* Oxford University Press, New York.

Rees, Martin, 1997, *Before the Beginning: Our Universe and Others,* Simon & Schuster, London.

Rees, Martin, 1999, *Just Six Numbers: The Deep Forces that Shape the Universe,* Weidenfeld & Nicolson, London.

Smolin, Lee, 1997, *The Life of the Cosmos,* Weidenfeld & Nicolson, London (Oxford University Press, New York).

Thorne, Kip, 1994, *Black Holes and Time Warps: Einstein's Outrageous Legacy,* Norton, New York.

Tipler, Frank, 1995, *The Physics of Immortality,* Doubleday, New York.

Weinberg, Steven, 1977, *The First Three Minutes,* Basic Books, New York (André Deutsch, London).

Weinberg, Steven, 1993, *Dreams of a Final Theory: The Search for the Fundamental Laws of Nature,* Vintage, London.

Wheeler, John Archibald, 1990, *A Journey into Gravity and Spacetime,* Scientific American Library, New York.

Will, Clifford, 1986, *Was Einstein Right?,* Basic Books, New York.

BIBLIOGRAPHY

Arnowitt, R., Deser, S. and Misner, C.W., 1962, 'The dynamics of general relativity', in *Gravitation: An Introduction to Current Research*, L. Witten (ed.), Wiley, New York.

Baierlein, R.F., Sharp, D.H. and Wheeler, J.A., 1962, 'Three-dimensional geometry as a carrier of information about time', *Physical Review*, **126**, 1864.

Banks, T., 1985, '*TCP*, quantum gravity, the cosmological constant and all that . . .', *Nuclear Physics*, **B249**, 332.

Barbour, J.B., 1989, *Absolute or Relative Motion?* Vol. 1: *The Discovery of Dynamics*, Cambridge University Press, Cambridge; reprinted as the paperback *The Discovery of Dynamics*, Oxford University Press, New York (2001).

Barbour, J.B., 1994a, 'The timelessness of quantum gravity: I. The evidence from the classical theory; II. The appearance of dynamics in static configurations', *Classical and Quantum Gravity*, **11**, 2853.

Barbour, J.B., 1994b, 'On the origin of structure in the universe', in *Philosophy, Mathematics and Modern Physics'*, E. Rudolph and I.-O. Stamatescu (eds), Springer, Berlin.

Barbour, J.B., 1999, 'The development of Machian themes in the twentieth century', in *The Arguments of Time*, J. Butterfield (ed.), published for The British Academy by Oxford University Press.

Barbour, J.B., 2001, 'General covariance and best matching', in *Physics Meets Philosophy at the Planck Length*, C. Callender and N. Huggett (eds), Cambridge University Press.

Barbour, J.B., forthcoming, *Absolute or Relative Motion?*, Vol. 2, Oxford University Press, New York (see note to Chapter 4 on p.344).

Barbour, J.B., Foster, Brendan Z. and Ó Murchadha, N., 2000, 'Relativity without relativity', http//xxx.lanl.gov/abs/gr-qc/0012089.

Barbour, J.B. and Ó Murchadha, N., 1999, 'Classical and quantum gravity on conformal superspace', http//xxx.lanl.gov/abs/gr-qc/9911071.

Barbour, J.B. and Pfister, H. (eds), 1995, *Mach's Principle: From Newton's Bucket to Quantum Gravity*, Birkhäuser, Boston.

Bell, J., 1987, *Speakable and Unspeakable in Quantum Mechanics*, Cambridge University Press, Cambridge.

Bohm, D., 1952, 'A suggested interpretation of the quantum theory in terms of "hidden" variables: I and II', *Physical Review*, **85**, 166 (reprinted in Wheeler and Zurek 1983).

Boltzmann, L., 1895, 'On certain questions of the theory of gases', *Nature*, **51**, 413.

Born, M., 1982, *Einstein's Theory of Relativity*, Dover, New York.

Brout, R., 1987, 'On the concept of time and the origin of the cosmological temperature', *Foundations of Physics*, **17**, 603.

Brown, H.R., 1996, 'Mindful of quantum possibilities', *British Journal for the Philosophy of Science*, **47**, 189.

Carnap, R., 1963, 'Autobiography', in *The Philosophy of Rudolf Carnap*, P.A. Schilpp (ed.), Library of Living Philosophers, P.A.

Clemence, G.M., 1957, 'Astronomical time', *Reviews of Modern Physics*, **29**, 2.

DeWitt, B.S., 1967, 'Quantum theory of gravity', *Physical Review*, **160**, 1113.

DeWitt, B.S., 1970, 'Spacetime as a sheaf of geodesics in superspace', in *Relativity*, M. Carmeli *et al.* (eds), Plenum, New York.

DeWitt, B.S. and Graham, N. (eds), 1973, *The Many-Worlds Interpretation of Quantum Mechanics*, Princeton University Press, Princeton.

Dirac, P.A.M., 1983, 'The evolution of the physicists' picture of nature', *Scientific American*, **208**, 45.

Eliot, T.S., 1964, *Knowledge and Experience in the Philosophy of F. H. Bradley*, Faber & Faber, London.

Fischer, A.E., 1970, 'The theory of superspace', in *Relativity*, M. Carmeli *et al.* (eds), Plenum, New York.

Giulini, D., 1999, 'The generalized thin-sandwich problem and its local solvability', *Journal of Mathematical Physics*, **40**, 2470.

Halliwell, J. and Hawking, S.W., 1985, 'Origin of structure of the universe', *Physical Review D*, **31**, 1777.

Halliwell, J.J., Pérez-Mercader, J. and Zurek, W.H. (eds), 1994, *The Physical Origins of Time Asymmetry*, Cambridge University Press, Cambridge.

Heisenberg, W., 1930, *The Physical Principles of the Quantum Theory*, Chicago University Press, Chicago.

Hofman, S.A., Kuchař, K. and Teitelboim, C., 1976, 'Geometrodynamics regained', *Annals of Physics*, **96**, 88.

Holland, P., 1993, *The Quantum Theory of Motion*, Cambridge University Press, Cambridge.

Isham, C., 1993, 'Canonical quantum gravity and the problem of time', in *Integrable Systems, Quantum Groups, and Quantum Field Theories*, L.A. Ibort and M.A. Rodriguez (eds), Kluwer Academic, London.

Kiefer, C., 1997, 'Does time exist at the most fundamental level?', in *Time, Temporality, Now*, H. Atmanspacher and E. Ruhnau (eds), Springer, Berlin.

Kuchař, K., 1992, 'Time and interpretations of quantum gravity', in *Proceedings of the 4th Canadian Conference on General Relativity and Relativistic Astrophysics*, G. Kunstatter *et al.* (eds), World Scientific, Singapore.

Lanczos, C., 1986, *The Variational Principles of Mechanics*, Dover, New York.

Lapchinskii, V.G. and Rubakov, V.A., 1979, 'Canonical quantization of gravity and quantum field theory in curved space-time', *Acta Physica Polonica* **B10**, 1041.

Liebscher, D.-E., 1999, *Einsteins Relativitätstheorie und die Geometrien der Ebene*, Teubner, Stuttgart.

Littlejohn, R.G. and Reinsch, M., 1997, 'Gauge fields in the separation of rotations and internal motions in the n-body problem', *Reviews of Modern Physics*, **69**, 213.

Mach, E., 1883, *Die Mechanik in ihrer Entwickelung. Historisch–kritisch dargestellt,* J.A. Barth, Leipzig. (English translation (1960): *The Science of Mechanics: A Critical and Historical Account of its Development,* Open Court, LaSalle, IL.)

McTaggart, J.M.E., 1927, *The Nature of Existence,* Cambridge University Press, Cambridge.

Misner, C.W., Thorne, K.S. and Wheeler, J.A., 1973, *Gravitation,* W.H. Freeman, San Francisco.

Page, D. and Wootters, W., 1983, 'Evolution without evolution: Dynamics described by stationary observables', *Physical Review D,* **27,** 2885.

Pais, A., 1982, *'Subtle is the Lord . . .' The Science and Life of Albert Einstein,* Oxford University Press, Oxford.

Pinker, S., 1997, *How the Mind Works,* Penguin, London/Norton, New York.

Poincaré, H., 1902, *La Science et l'hypothèse,* Flammarion, Paris. (English translation (1905): *Science and Hypothesis,* Walter Scott, London.)

Poincaré, H., 1904, *La Valeur de la Science,* Flammarion, Paris. (English translation (1907): *The Value of Science,* Science Press, New York; reprinted in 1958 by Dover, New York.)

Schrödinger, E., 1944, *What is Life?,* Cambridge University Press, Cambridge.

Shimony, A., 1997, 'Implications of transience for spacetime structure', in *The Geometric Universe: Science, Geometry, and the Work of Roger Penrose,* S.A. Huggett *et al.* (eds), Oxford University Press, Oxford.

Silk, J., 1994, *A Short History of the Universe,* W.H. Freeman/Scientific American Library, New York, p. 87.

Smolin, L., 1991, 'Space and time in the quantum universe', in *Conceptual Problems of Quantum Gravity,* A. Ashtekar and J. Stachel (eds), Birkhäuser, Boston.

Smolin, L., 2001, *Three Roads to Quantum Gravity,* Weidenfeld & Nicolson, London (Basic Books, New York).

Stachel, J. (ed.), 1987, *The Collected Papers of Albert Einstein,* Vol. 1, *The Early Years, 1879–1902,* Princeton University Press, Princeton.

Tait, P.G., 1883, 'Note on reference frames', *Proceedings of the Royal Society of Edinburgh,* Session 1883–4, p. 743.

Vilenkin, A., 1989, 'Interpretation of the wave function of the Universe', *Physical Review D,* **39,** 1116.

Walsch, N.D., 1997, *Conversations with God—An Uncommon Dialogue,* Hodder & Stoughton, London.

Wheeler, J.A., 1964a, 'Mach's principle as boundary condition for Einstein's equations', in *Gravitation and Relativity,* H.-Y. Chiu and W.F. Hofmann (eds), Benjamin, New York.

Wheeler, J.A., 1964b, 'Geometrodynamics and the issue of the final state', in *Relativity, Groups, and Topology: 1963 Les Houches Lectures,* C. DeWitt and B. DeWitt (eds), Gordon & Breach, New York.

Wheeler, J.A. and Zurek, W. (eds), 1983, *Quantum Theory and Measurement,* Princeton University Press, Princeton.

Williams, T., 1951, 'The timeless world of a play', in *The Rose Tattoo,* New Directions Books, New York.

Zeh, H.-D., 1992, *The Physical Basis of the Direction of Time,* 2nd edn, Springer, Berlin (3rd edn, 1999).

INDEX